Forest Policy, Economics, and Markets in Zambia

Forest Policy, Economics, and Markets in Zambia

Edited by

Phillimon Ng'andwe, Jacob Mwitwa and Ambayeba Muimba-Kankolongo
School of Natural Resources,
Copperbelt University,
Kitwe, Zambia

AMSTERDAM • BOSTON • HEIDELBERG • LONDON
NEW YORK • OXFORD • PARIS • SAN DIEGO
SAN FRANCISCO • SINGAPORE • SYDNEY • TOKYO
Academic Press is an imprint of Elsevier

Academic Press is an imprint of Elsevier
125, London Wall, EC2Y 5AS
525 B Street, Suite 1800, San Diego, CA 92101-4495, USA
225 Wyman Street, Waltham, MA 02451, USA
The Boulevard, Langford Lane, Kidlington, Oxford OX5 1GB, UK

Notices
Knowledge and best practice in this field are constantly changing. As new research and
experience broaden our understanding, changes in research methods or professional practices,
may become necessary.

Practitioners and researchers must always rely on their own experience and knowledge in
evaluating and using any information or methods described herein. In using such information or
methods they should be mindful of their own safety and the safety of others, including parties for
whom they have a professional responsibility.

To the fullest extent of the law, neither the Publisher nor the authors, contributors, or editors,
assume any liability for any injury and/or damage to persons or property as a matter of products
liability, negligence or otherwise, or from any use or operation of any methods, products,
instructions, or ideas contained in the material herein.

ISBN: 978-0-12-804090-4

British Library Cataloguing-in-Publication Data
A catalogue record for this book is available from the British Library

Library of Congress Cataloging-in-Publication Data
A catalog record for this book is available from the Library of Congress

For Information on all Academic Press publications
visit our website at http://store.elsevier.com/

CONTENTS

Chapter 1 An Overview of the Forestry Sector in Zambia
Phillimon Ng'andwe, Jacob Mwitwa, Ambayeba Muimba-Kankolongo, and Jegatheswaran Ratnasingam

Chapter 2 Wood and Wood Products, Markets and Trade
Phillimon Ng'andwe, Jegatheswaran Ratnasingam, Jacob Mwitwa, and James C. Tembo

Chapter 3 Non-Wood Forest Products, Markets, and Trade

Ambayeba Muimba-Kankolongo, Phillimon Ng'andwe,
Jacob Mwitwa, and Mathew K. Banda

Chapter 4 Contribution of the Forestry Sector to the National Economy

Phillimon Ng'andwe, Jacob Mwitwa, Ambayeba
Muimba-Kankolongo, Nkandu Kabibwa, and
Litia Simbangala

Chapter 5 Integration of Forestry into the National Economy
        Jacob Mwitwa, Ambayeba Muimba-Kankolongo,
        Obote Shakacite, and Phillimon Ng'andwe

LIST OF CONTRIBUTORS

Mathew K. Banda
Department of Computer Science, School of Mathematics and Natural Sciences, Copperbelt University, Kitwe, Zambia

Nkandu Kabibwa
Central Statistical Office, Lusaka, Zambia

Ambayeba Muimba-Kankolongo
Department of Plant and Environmental Sciences, School of Natural Resources, Copperbelt University, Kitwe, Zambia

Jacob Mwitwa
Department of Plant and Environmental Sciences, School of Natural Resources, Copperbelt University, Kitwe, Zambia

Phillimon Ng'andwe
Department of Biomaterials Science and Technology, School of Natural Resources, Copperbelt University, Kitwe, Zambia

Jegatheswaran Ratnasingam
Faculty of Forestry, University Putra Malaysia, Serdang, Selangor, Malaysia

Obote Shakacite
Department of Plant and Environmental Sciences, School of Natural Resources, Copperbelt University, Kitwe, Zambia

Litia Simbangala
Central Statistical Office, Lusaka, Zambia

James C. Tembo
Department of Biomaterials Science and Technology, School of Natural Resources, Copperbelt University, Kitwe, Zambia

FOREWORD

It is widely and generally recognized that the value of forest for life, and the potential benefits and solutions they bring to mankind are enormous. Forests sustain livelihoods, host biodiversity, help stabilize the climate, provide sustainable wood materials, contribute to greening the environment, protect water sources, and provide foods, medicines and many other non-wood forest products. At the same time, forests are not only under threat from deforestation and forest degradation, but also from fires, pests, and diseases. Historically, how people interact with forests has been a reflection of our complex and at times contentious relationship with our natural environment. Forests contribute directly and indirectly to poverty reduction by providing jobs and incomes for rural communities in the country. It is also clear that despite different classifications of forest vegetation in the country, a common phenomenon has been noticed: The communities' dependence on the forest for their survival.

Many writers in the past have recognized forests as being valued largely for their environmental benefits. However, the authors of this book have endeavored to demonstrate the equally important role of both wood and non-wood forest products in providing economic, social, and cultural benefits to the people in many of their geographical locations in the country. The researcher's findings which comprise this publication draw upon diverse experiences from around the country, reflecting on the many ways in which people deal with production, marketing, and trading in wood and non-wood forest products.

This publication is not a simple amalgam of writings and articles on the subject. It is virtually a complete, new and well thought out presentation which depicts the correct realities of people's interaction with the forests. This first edition has been written particularly with the aim of providing insightful information to those who study forestry, those who sustain their livelihoods through interaction with the forest and those who come across wood and non-wood forest products in pursuit of social, economic, and cultural endeavors.

The book has provided a concise and precise overview of the Forestry sector in Zambia and exposes the inadequacies in the forest policies and lacunae in the legal framework, which in most cases, appear archaic and out of touch with emerging issues in the sector, such as climate change, carbon management and trade, forest and timber certification, among others. However, given the life and daily endeavors of mankind, it is not every man's action or social interaction with nature which could be regulated by formalized or enacted laws such as the Forests Act and its related subsidiary legislation. Achieving sustainable forest management solutions ultimately require a process of dialogue and shared learning. This publication contributes to such a process by bringing together the challenges faced in the management of forests and the best practices from the Government and rural communities.

The learned authors of this publication have articulately blended the forest activities and strategies in both natural and exotic forest plantations. The contribution of industrial plantations, which are largely managed by the Zambia Forestry and Forest Industries Corporation Limited (ZAFFICO), to the country's wood industry, cannot be overemphasized. ZAFFICO's strategy has been revised and it is tailored towards enhancement of annual replanting and expansion of forest plantations in other areas outside the Copperbelt Province in order to ensure sustainable wood supply in the country. One of the biggest challenges of the future lies in the area of tension to be found between increasing demand for timber on the one hand and effective conservation of forests and nature on the other.

The daunting task in both natural and exotic forest plantations, however, is how to deal with markets and trade in timber, which is characterized by unprocessed products, primarily round wood, which entails that the full value of wood forest resources is not adequately realised. Improved skill, machinery and capital are often critical layers of technological development. Due to lack of value addition, most of the wood in natural forests is used for woodfuel. Even valuable species of wood are burnt and wasted. These wood resources, if well harvested, would contribute significantly to the country's revenue and contribute to economic development.

Lack of statistics on the forestry sector in general and trade and markets for wood and non-wood forest products has been cited in the book as a challenge, but the authors have strived to make immeasurable effort to display and depict the near exact scenario in the country. Also the comparative and illustrative approach in this publication is intrinsically connected to the realities on the ground.

In my career as a Legal practitioner, and former Commissioner of Lands for Republic of Zambia, I have come across a wide variety of books and other sources designed to give an understanding of land tenure issues and forest management in Zambia. However, this publication is one of the most comprehensive collections of hands-on-information I have come across. For the workplace and teaching environment, it is potentially one of the most useful, because here, the learned authors have gathered together, into one succinct, handy volume, an amazing wealth of information on the status of wood and non-wood forest products, strategies on how to manage and exploit them, expertise, advice, and solutions. The authors have also presented, in a simple, straight forward fashion, the core information as cross—referenced solutions, and then complimented these valuable authorities through tables and thereby making it easier for the readers to access them, but without the sacrifice of precision and accuracy of information.

A common understanding is that there is no single accepted forest management standard worldwide, and each system takes a somewhat different approach in defining standards for sustainable forest management. The suggestions and recommendations made in this book provide appropriate measures and solutions which the Government, the private sector, Nongovernmental Organizations and cooperating partners will find valuable in the practice and management of forestry and all its products in the country. Therefore, in a country like ours where the challenges are many—from growing unemployment to deforestation—the need to manage our forest resources sustainably has become a matter of urgency. This publication may prompt a change in the casual approach in which we deal with wood and non-wood forest products that we may have taken for granted for too long.

To all members of the academia, researchers, and forest practitioners, we all need to build on the momentum and goodwill demonstrated through this book, and to carry the message of preserving, protecting, and participating in all spheres of trade, value addition and marketing of wood and non-wood forest products for the benefit of current and future generations in the country.

Frightone Sichone
Managing Director
Zambia Forestry and Forest Industries
Corporation Limited, Ndola, Zambia

The continuous over-exploitation of forests and natural resources has led to considerable degradation of large land areas as well as to depletion of their rich biological diversity, which constitute the livelihood support system for the majority of the households especially in rural areas. Moreover, the forests and forestry goods and services constitute an ideal component of many country's natural resources that could serve as a potential tool for national development. About more than half of the population of Zambia and of several other countries in sub-Saharan Africa, more especially in the rural area, derive their livelihoods directly from forests. Despite the concern over continuous forest misuse, there has been little consideration on how forest resources could be utilized in a sustainable manner to generate ecological, environmental, and socioeconomic benefits to the population. In addition, efforts to integrate forestry-related goods and services into the development planning process have not been existent. Yet, the benefits that could accrue from exploitation of the forest natural resources by the local people are numerous and varied. Concerns have been therefore raised by different stakeholders about the inclusion of the forestry sector in national accounting processes, such as is done in the manufacturing or agriculture sectors. The book enlightens the effective contribution of the forestry sector to the population's livelihoods through improved collection of forestry statistics that foster the understanding and integration of the forestry sector in poverty reduction processes and the national economy to enhance its integration in national planning.

The Forest Policy, Economics and Markets text book is the first volume describing the wood and non-wood forest products in Zambia. It is an all-in-one information package, a one stop reference book about the characteristics of the forest sector that provides information on production, market, and trade as well as related policies and legislations. This volume is a compilation from research and surveys conducted in various provinces and districts of Zambia. In the book, the production and various uses of wood and non-wood forest products in different parts of Zambia are presented in a way that enables the reader to appreciate this sector in detail.

The book is organized in five chapters written by experts in the field as follows:

In **Chapter 1**, a forestry sector definition has been given in relation to what it constitutes, as an overview, based on the International Standard Industry Classification of economic activities, policies and legal frameworks. Production from forest and logging has been split into: (a) Natural forests, and (b) forest plantations and highlighting the agro-ecological zones in which they are found. Forest and forest plantation productivity and round wood supply potential are presented in Tables and Figures. Emerging issues, such as forest certification, green building and UN-REDD, have been included to enable the reader to appreciate the global and national efforts aimed at developing appropriate policies and frameworks.

In **Chapter 2**, authors have focused on production, markets and trade of wood and wood products in Zambia. The manufacture of wood and wood products, pulp and paperboard products since 2001 are described to show industry development and areas of growth potential. The chapter identifies value addition to wood and wood products, including energy production from forest and mill waste, as an important basis for growth and job creation in the sector.

In **Chapter 3**, the authors review the role of non-wood forest products (NWFP), particularly various edible wild roots, tubers and bulbs, production and trade. The authors explore linkages of NWFP to the livelihood of the population in urban and rural areas including their contribution to poverty reduction.

Chapter 4 highlights the contribution of the forestry sector to Zambia's economy. Production and trade from key economic activities including forest and logging, manufacture of wood and wood products, as well as paper and paperboard products, spanning over a period of 10 years are presented in tables and figures.

Chapter 5 summarizes efforts to integrate the forestry sector into the national planning process using the existing policies, legal frameworks, and procedures. The linkage of the forestry sector to the national economy is summarized in matrix form that will enable readers to see the sector insights.

The book reflects a detailed overview of the Forest Policy, Economics, and Markets in relation to the contribution of wood and non-wood forest products to the national economy and poverty alleviation in Zambia. It is intended for academic institutions, policy makers, entrepreneurs, researchers, and for other groups with global and local special interest in the forestry sector.

REFERENCES

The main objective of the references given is to guide readers to additional information. It is not intended to be complete or exhaustive.

CHAPTER 1

An Overview of the Forestry Sector in Zambia

Phillimon Ng'andwe[a], Jacob Mwitwa[b], Ambayeba Muimba-Kankolongo[b], and Jegatheswaran Ratnasingam[c]

[a]Department of Biomaterials Science and Technology, School of Natural Resources, Copperbelt University, Kitwe, Zambia; [b]Department of Plant and Environmental Sciences, School of Natural Resources, Copperbelt University, Kitwe, Zambia; [c]Faculty of Forestry, University Putra Malaysia, Serdang, Selangor, Malaysia

FORESTRY SECTOR CLASSIFICATION

The forestry sector in Zambia embraces all economic activities defined in the International Standard Industry Classification (ISIC Revision 4) (UN, 2006a; Ng'andwe et al., 2012) that are essential for the country to develop. These are:

- *Forest and logging* which includes various economic activities taking place in forests and woodlands such as silviculture, production of round-wood, extraction and gathering of non-wood forest products (NWFPs) as well as activities on products that undergo little processing such as firewood, charcoal, pit-props, slab poles, and utility poles.
- *Manufacturing of wood and products of wood* comprising production of sawnwood, wood-based panels, builders joinery, and carpentry.
- *Pulp, paper, and paperboard products.*

The forestry sector of Zambia, is therefore, not limited to forestry activities, as a boundary, but includes the downstream processing activities involving wood and NWFPs. The contribution of the sector to the national economy, forest policies, legal frameworks and sector's integration are considered cross-cutting these boundaries in this book.

Forest Policies and Legal Frameworks

The Forestry Department is mandated by the Zambian government to manage resources in forest and customary lands other than leasehold (GRZ, 1965, 1973, 1998b). This mandate is supported by the Forest Policy of 1965 and the Forest Act No. 39 of 1973. The forestry legislation was established in 1949 during the colonial administration and revised in 1960

Forest Policy, Economics, and Markets in Zambia. DOI: http://dx.doi.org/10.1016/B978-0-12-804090-4.00001-X

and 1970 to Act No. 39 of 1973 (IDLO, 2011). The 1973 Act has been the active legal instrument that provides for a centralized forest management, with government having absolute power over all aspects of forest and woodland management. The inadequacy of this legal instrument to address emerging issues resulted in the development of the Forest Act of 1999 (GRZ, 1999) which provides for the formation of the Zambia Forestry Commission, a semi-autonomous body. The Joint Forest Management initiative was started under the 1999 and 2006 scrics of statutory instruments derived from the inactive Forest Act of 1999 as a model for involving communities in forest management and utilization programs in Zambia.

The Zambia Forest Action Plan (ZFAP) of 1997 (ZFAP, 1997), Forest Policy of 1998 and Forest Act of 1999 (GRZ, 1999) provide for industry and community participation in forest management and utilization. Formulated under the National Environmental Action Plan (NEAP) of 1994, the Forest Policy of 1998 is aimed at changing the national institutional and legal framework for forest management and administration so that the forestry sector could be administered by a forest commission. There are also cross-cutting legal frameworks such as the Lands Act of 1995 (GRZ, 1995) and the draft Lands Policy of 1998 (GRZ, 1998a) which legally distinguish property rights tied to forests and those tied to land (IDLO, 2011). In addition, the Investment Act of 1993 provides a legal framework for investment in any sector of the economy that includes forestry.

ZFAP established a framework for planning in forestry and contributes to the preparation of forest policies, action plans and programs. It raises awareness of issues related to the forestry sector and provides for updates of forest policy. The National Forest Policy (NFP) of 1998 addresses sustainable management and utilization of forest resources using broad-based approaches, such as:

i. resource management and development;
ii. resources allocation;
iii. capacity building; and
iv. gender equality.

Under the resource management and development, both ZFAP and NFP aim to meet on a sustainable basis the demand for woodfuel, timber, poles and NWFPs, to contribute to the national economy by creating formal and informal employment in wood and NWFPs processing and to conserve forest ecosystems and biodiversity.

The lack of full implementation of the policy and legal frameworks for the forestry sector indicates the inadequacies in the Forestry Department institutional capacity to carry out its mandate effectively and its weakness in addressing emerging challenges such as climate change issues, decentralization, benefit sharing, and Private Public Partnership (GRZ, 2010a; IDLO, 2011; Mwitwa and Makano, 2012). However, the government of Zambia has had aspirations to improve the forestry sector (GRZ, 2006, 2011a), but the institution capacity to fully conduct its mandates provided in the legal and fiscal frameworks has not been strengthened due to delays in enacting such instruments by Parliament.

Forest Classes and Agro-Ecological Regions
The Forest Act No. 39 of 1973 legally subdivides natural forests into four land classes: Trust, Reserve, and State lands, as well as lands with designated forest functions. Forests on reserve lands are classified either as national or local forests and any activity to be carried out in these forests are regulated under the use rights obtained through licenses. According to the legal classification, national forests are identified for industrial processing for national benefits while local forests are for community interests. Within this broad classification, in which accessibility for exploitation is regulated, there are also protected forests and forests on customary land (GRZ, 1997; Chileshe, 2001) as well as global agro-ecological zones (AEZs). Zambia has three main different AEZs described in Integrated Land Use Assessment (ILUA) Report of 2008 I (MTENR, 2008b) and corresponding to various forest biomass stocks, described below (Chidumayo, 2012).

- *AEZ I—Luangwa—Zambezi river zone*: This zone is found in the dry Miombo woodland of Zambia and covers Western, Eastern, Lusaka and Southern Provinces. It spans over 8.3 million ha and is characterized by low rainfall of less than 800 mm/year. The tree production potential is low. The above-ground wood biomass production is estimated at 6.1 million tons per annum (Chidumayo, 1996a) and is co-dominated by *Brachystegia spiciformis Julbernadia paniculata*, *Burkea africana*, and *Diplorhynchus condylocarpon*, as common understory taxa (Chidumayo, 2012).
- *AEZ IIa—Central, Southern and Eastern Plateau*: This region covers over 11.7 million ha with rainfall ranging between 800 and 1,000 mm/year and is found in Central, Eastern, Lusaka, Northern, Northwestern and parts of Western Provinces. It is characterized by

high agriculture and forest production potential. The above-ground wood biomass production is estimated s estimated at 8.4 million tons per annum (Chidumayo, 2012). The Central drier Miombo woodland have *Brachystegia* spp. and *Julbernardia globiflora* as co-dominant species while *D. condylocarpon, Lannea* spp., *Ochna* spp., and *Pseudolachnostylis lismaprouneifolia* are common understory species. In the Eastern drier Miombo woodland, *Brachystegia manga and Julbernardia* spp. woodlands are dominant with *Diospyros* spp., *D. condylocarpon, Ochna* spp., and *P. maprouneifolia* as common understory taxa (Chidumayo, 2012).

- *AEZ IIb—Western semi arid plains*: This agro-ecological zone covers 5.2 million ha and is characterized by low rainfall compared to AEZ IIa, low agriculture and low tree production potential while wood biomass production is estimated at 3.9 million tons per annum. As reported by Chidumayo (2012) this AEZ corresponds to Western drier Miombo which is co-dominated by *Baikiaea plurijuga, B. spiciformis, J. paniculata, B. africana,* and *D. condylocarpon* as common understory taxonomic groups. This is one of the most active areas located in the high conservation value forest and for production of sawn timber from Zambezi teak(*B. plurijuga*) as far back as 1911 (Figure 1.1). Today, *B. plurijuga*, continues to be most exploited by the sawmill industry and is the main economic activity in this agro-economic zone.
- *AEZ III—Northern, Copperbelt and Northwestern high rainfall (1,000–1,500 mm per annum)*: This AEZ covers over 24.6 million ha and is more than half of the forest area of Zambia. It is characterized by high production of both agriculture crops and forestry with the above-ground biomass production estimated at 57.2 million tons per annum. This zone is found in northern wet Miombo (Chidumayo, 2012) and is

(a)   (b)   (c)

Figure 1.1 The Zambezi teak forest timber production value chain. (a) Zambezi teak natural forest, (b) sawlogs, and (c) sawn wood.

dominated by *Brachystegia* spp., *J. paniculata*, and *Parinari curatellifolia* as common co-dominant canopy, and *Monotes africanus*, *Syzygium guineense macrocarpum*, and *Uapaca* spp. as common understory species. In Northwestern wetter Miombo, *Brachystegia* spp., *Isoberlinia angolensis*, and *J. paniculata* are co-dominants. Other common understory species include *Anisophyllea boehmii*, *D. condylocarpon*, *S. guineense*, *Macrocarpum* spp., and *Uapaca* spp.

Classification of Forest Vegetation
The forests of Zambia constitute one of the largest natural resources in sub-Saharan Africa covering approximately 66% of the total land area in various categories of vegetation type. The first category comprises the closed forests consisting of the evergreen forests, dry deciduous forests, montane forests, swamp forests and riparian forests. These forest types constitute different vegetation types with the Miombo woodland being the most extensive and covering about 47% of the country. The Miombo woodland is characterized by the presence of numerous species of *Brachystegia, Julbernadia*, and *Isoberlinia* spp. (Campbell, 1996). The Miombo woodlands are economically important vegetation type for supply of timber, poles, firewood and charcoal. It is also the source of many NWFPs, such as honey, medicine, mushrooms, and caterpillars.

In the second category are the open forests, commonly known as the savannah woodlands, which account for 75% of the total land area in Zambia. The most dominant in this category is the Kalahari woodland vegetation—found on Kalahari sands in Western Zambia. This vegetation type is dominated by *Guibourtia, Burkea, Brachystegia, Isoberlinia, Julbernadia*, and *Schinziophyton* species. The third and fourth woodlands are the Mopane and Munga woodlands characterized by vegetation of the *Colophospermum mopane* and *Acacia* species. The grassland vegetation, which includes wetlands (flood plains or swamps) and dambos, is found around ephemeral rivers and covers about 17% of the land area of Zambia ranging from pure grasslands to grasslands with scattered trees (Chidumayo, 1996a,b).

The termitaria forests are found scattered throughout Zambia wherever the soil is not sandy. General biological activity is greater and organic matter decomposes faster on mound soils because decomposers are more numerous. Termitaria occurs more frequently on dambo margins, Munga and Mopane woodlands than in Miombo woodlands and are usually scarce in dry evergreen forests.

Forest and Woodland Productivity

The potential for production of wood and NWFPs is quite large from the semi-evergreen forests, corresponding to the wet Miombo woodlands which are widely distributed throughout Zambia. The semi-evergreen forests have the largest commercial forest growing stock (CFGS) and annual allowed cut (AAC) estimated over a total of 256 million m^3 and 6.69 million m^3 per annum, respectively, seconded by the deciduous forests with 72.6 million m^3 and AAC of 2.18 million m^3 (MTENR, 2008b) (Table 1.1).

It is estimated that 57% of Zambia's natural forests fall under the wet Miombo woodland and account for over 70% of the commercial growing stock (MTENR, 2008b) based on the normalized difference vegetation index (NDVI). The NDVI is a simple graphical indicator that can be used to analyze remote sensing measurements, typically but not necessarily from a space platform, and assess whether the target being observed contains live green vegetation or not and is an indicator of degradation (Meneses-Tovar, 2011). Thus, NDVI was one of the most successful of many attempts to simply and quickly identify vegetated areas and their "condition," and it remains the most well-known and used index to detect live green plant canopies in multispectral remote sensing (Chidumayo, 2012). Negative values of NDVI (values approaching −1) correspond to water. Values close to zero

Table 1.1 The Forest Growing Stock in Natural Forests, Woodlands and Forest Plantations of Zambia

Biome		Forest Growing Stock		Commercial Forest Growing Stock		
		million ha	million m^3	million m^3	MAI (m^3/ha/a)	AAC (million m^3/a)
Natural forests	Evergreen	0.820	54.800	10.2000	2.000	0.300
	Semi-evergreen	34.15	2,127.000	256.000	1.600	6.690
	Deciduous	14.87	595.400	72.600	1.200	2.180
	Other	0.140	164.000	26.900	–	–
	Subtotal	**49.97**	**2,941.200**	**365.700**	**1.610**	**9.650**
Plantations	Hardwoods	0.011	2.240	2.000	7.000	0.080
	Softwoods	0.046	6.970	5.300	4.000	0.180
	Subtotal	**0.057**	**9.210**	**7.300**	**5.500**	**0.281**
Grand total		**50.030**	**2,950.400**	**373.150**		**9.930**

Note: MAI is the mean annual increment in cubic meters per hectare per annum (m^3/ha/a), AAC is the annual allowable cut in cubic meters per annum (m^3/a).

(−0.1 to 0.1) generally correspond to barren areas of rock, sand, or snow and low, positive values represent shrub and grassland (approximately 0.2−0.4). High values (approaching 1) indicate temperate and tropical rainforests (www.fao.org/giews).

Chidumayo (2012) reported that the wet Miombo woodland is associated with high levels of productivity throughout the year ranging from NDVI values of 175−285 between November and April and lower NDVI values in the dry season, reaching 160 in September. The mean annual increment (MAI) ranges from 1.2 m³/ha/year in the deciduous forests to 2.0 m³/ha/year in the tropical moist evergreen forests (Chidumayo, 1996a; Chisanga, 2005). There are also other forest types including evergreen, deciduous forests, and woodlands in Zambia from which wood and NWFPs are derived. The annual allowable cut is estimated at 0.3 million m³ per year from the evergreen, 6.69 million m³ in the semi-evergreen, and 2.18 million m³ in deciduous forests, respectively (Table 1.1). According to Chidumayo (2012) the main floristic association of commercial importance from different phenologies include *E. delevoyi, Brachystegia* spp., *Pterocarpus Angolensis, B. plurijuga,* and *Guibourtia* spp. (Figure 1.1). The *Brachystegia* species—characterized by non-durability and processing difficulties—are dominant in the semi-evergreen and deciduous forest types.

Chidumayo (2012) reported that the wet Miombo woodland is associated with high levels of productivity throughout the year ranging from NDVI values of 175−285 between November and April and lower NDVI values in the dry season, reaching a low of about 160 in September. The MAI ranges from 1.2 m³/ha/year in the deciduous forests to 2.0 m³/ha/year in the tropical moist evergreen forests (Chidumayo, 1996a; Chisanga, 2005). There are also other forest types including evergreen, deciduous forests and woodlands in Zambia from which wood and NWFPs are derived. The annual allowable cut is estimated at 0.3 million m³ per year from the evergreen, 6.69 million m³ in the semi-evergreen and 2.18 million m³ in deciduous forests, respectively (Table 1.1). The main floristic association of commercial importance as reported by Chidumayo (2012) include *E. delevoyi, Brachystegia* spp., *P. angolensis, B. plurijuga,* and *Guibourtia* spp.

Forest Plantations Productivity
At a global level, the plantation modeling study by Brown (1999) estimates that forest plantations in 1995 had the potential to produce 331 million m³ or 22.2% of global industrial round-wood production.

The Global Forest Products Model used by Brown (1999) and Whiteman and Brown (1999) suggests industrial round-wood production in Africa region to be around 84 million m^3 by 2010, with a round-wood equivalent consumption estimate of 37 million m^3. The proportion of global industrial round-wood that could be sourced from forest plantations was projected to increase to between 30.6% and 46.6% by 2020 (Brown, 1999).

According to Ng'andwe (2012a), Zambia's plantation wood production contributes an insignificant fraction to global industrial round-wood production and of potential global production from forest plantations because forest plantation expansion stopped during the 1990s. The annual allowable cut from forest plantations dropped from 0.614 million m^3 per annum in 2001 to 0.281 million in 2010 (ZAFFICO, 2007) (Table 1.1). Forestry plantations in Zambia play a critical role in the national economy; therefore, a decline in the supply potential negatively impacts on the development of the timber industry. Under the existing planting plans (ZAFFICO, 2007; GRZ, 2011), Zambia, in general, has not taken advantage of its fast tree growth potential to produce and contribute significant quantities of the world's plantation fiber production even to the domestic and southern African region markets. The establishment of forest plantation is currently about 2,500 ha/year and is primarily by ZAFFICO in the Copperbelt Province only. This is expected to change when plans for additional plantation establishment in Luapula and Muchinga Provinces by ZAFFICO are implemented.

Forest and Woodland Resource Trends

There have been several forest assessments and inventories dating as far back as 1932, 1942, 1952, 1984, 1999, 2002, and 2005 as reviewed in ILUA report (ILUA I) (MTENR, 2008b) which provides statistics on the forest growing stocks (FGSs) from natural forests (MTENR, 2008b) and forest plantations (ZAFFICO, 2004, 2007, 2014). These assessments differed in terms of accuracy of estimates and coverage but the focus in all has been the determination of Zambia's FGS available for wood production commercially (FAO, 2010). In this book the commercial growing stock refers to the volume of commercial timber species that include *P. angolensis, B. plurijuga, Afzelia quanzensis*, and *Guibourtia coleosperma* from natural forests and woodlands where harvesting of timber is permitted. It also refers to plantation grown timber species such as *Pinus kesiya, Pinus oocarpa, Eucalyptus grandis*, and *Eucalyptus cloeziana*.

In recognition of the lack of sound and reliable national level forest resource information, the Government of Zambia decided in 2005 to initiate the ILUA (MTENR, 2008a). The ILUA was based on a standard national forest assessment (NFA) approach developed by FAO, which has been applied in several other countries since 2000 (e.g., Costa Rica, Guatemala, Honduras, Lebanon, Cameroon, the Philippines, Bangladesh, and Nicaragua). The NFA design has been developed to ensure collection of a holistic set of data to meet a number of national and international information requirements. Elaborate description of the NFA methodology is available at the webpage of FAO's program for National Forest Monitoring and Assessment (NFMA) (www.fao.org/forestry/nfma).

As a comparison, the FRA 2005 reported the annual deforestation in Zambia to be around 444,800 ha per annum while the estimation conducted under the ILUA in 2008 suggested deforestation arising from land-use change to be in the range of 250,000 and 300,000 ha per annum. Conclusively, analysis of historical data sets (ZFAP, 1997; MTENR, 2008b) provides two estimates of FGS decline. Inventories reveal an altered standing volume per hectare from 94 m^3/ha in 1996, 83 m^3/ha in 2003, and 55 m^3/ha in 2006, but the ILUA report based on remote sensing data shows an increase in the volume of the FGS with the inclusion of resources from other forest types. The past estimates were since adjusted in 2009 based on the ILUA report which is expected to be verified in ILUA II (GRZ, 2010b; Kewin, 2009).

According to the ILUA I (MTENR, 2008b) the total FGS in Zambia increased from 1,461 million m^3 in 2001 to 2,910 million m^3 in 2010 based on the actual inventory of 2005–2008. The commercial growing stock is estimated at 365.6 million m^3 representing 12% of the FGS (MTENR, 2008b) as shown in Table 1.2.

There is a variation in the size of FGS and annual allowable cut of round-wood from various AEZs by Province with Northwestern accounting for a largest area.

The highest FGS is found in Northwestern Province with over 904 million m^3, followed by Central Province with over 485 million m^3 representing 31% and 17% of the country's total, respectively (MTENR, 2008b) (Table 1.3). Lusaka Province has the lowest FGS estimated at 88 million m^3.

Table 1.2 The Trends of Growing Stock from Natural Forests in Zambia (2001–2010)						
Description	UM	2001	2003	2005	2008	2010
Forest growing stock	million m^3	1,461	1,431	1,400	2,940	2,910
– Forests	million m^3	1,362	1,335	1,307	2,785	2,755
– Woodlands	million m^3	99	96	93	58.0	58
– Other	million m^3	–	–	–	97.0	97.0
Commercial FGS	million m^3	175.3	171.7	168.0	365.8	365.8
Evergreen forests (E)	million m^3	4.9	4.8	4.7	10.2	10.2
Semi-evergreen forests (E)	million m^3	175.3	120.2	117.6	256.0	256.0
Deciduous forests (E)	million m^3	34.80	34.08	33.34	72.6	72.6
other (E)	million m^3	12.92	12.65	12.38	26.95	27.0

Note: (E) indicates estimates; FGS is the forest growing stock.

Table 1.3 Area and Standing Timber Volume from Provinces of Zambia				
Province	Area (ha)	Standing Timber Volume (million m^3)		
		FGS	CFGS	AAC
Central	7.910	485.900	47.700	1.424
Copperbelt	1.609	173.300	23.400	0.386
Eastern	5.152	264.800	31.300	0.927
Luapula	3.465	158.300	17.500	0.832
Lusaka	1.651	88.000	5.200	0.198
Muchinga	3.000	170.000	21.000	0.800
Northern	4.212	175.000	21.900	0.930
Northwestern	10.041	904.000	116.000	2.410
Southern	4.672	135.200	17.300	0.751
Western	8.254	385.000	64.500	0.990
Total Zambia	**49.968**	**2,939.500**	**365.800**	**9.650**

Note: FGS is the forest growing stock, CFGS is the commercial forest growing stock and AAC is the annual allowable cut.

The FGS volume was adjusted upwards in 2008 following the integrated land-use assessment and inventory of 2005–2008. The figure was upwards to 2,939 million m^3 in 2010 (Table 1.3) (MTENR, 2008b). MTENR (2008b) estimated that about 12% of FGS (365 million m^3) is available for industrial and household timber consumption as CFGS. A total of 2,939.5 million m^3 of FGS volume from 49,968 million ha of natural forests and woodlands was estimated to have a potential to supply over 9.65 million m^3 of round-wood (MTENR, 2008b; Ratnasingam and Ng'andwe, 2012).

Forest and Logging Production Trends

The area designated for production in ILUA I report was estimated at 23.7% of the total natural forest area and an average MAI of 1.61 m³/ha/year with 12% of the volume being for commercial growing stock (MTENR, 2008b). The annual allowable cut is the wood that is available for industrial processing and that which can be sustainably supplied using the 19 listed timber species by the Forestry Department (MTENR, 2008b). The lesser used or known species will have to be promoted through research and development to determine the technical properties affecting their use. On the other hand, the area of CFGS from plantations is estimated at 9% (ZAFFICO, 2007) of the total FGS area, tabulated in matrix form (Table 1.4).

Many of the existing forest plantations of Zambia were established particularly in AEZ II and III. These zones are characterized by medium to high rainfall. The Copperbelt Province has the largest proportion of

Table 1.4 Matrix of the Provincial Round-Wood Production from Different Agro-Ecological Zones and Size of Forests in 2010						
Agro-Ecological Zone >	AEZ II ('000 ha)	AEZ I ('000 ha)	AEZ II a ('000 ha)	AEZ II b ('000 ha)	AEZ III ('000 ha)	Total Area (ha)
Rainfall >	*800–1,000*	*<800*	*800–1,000*	*Low*	*>1,000*	
Province	*Area 1,000 ha*					
Central	0.24	1,215	2,987	–	3,708	7,910
Copperbelt	52.00	–	–	–	1,609	1,609
Eastern	2.84	2,118	3,034	–	–	5,152
Luapula	0.89	–	–	–	3,465	3,465
Lusaka	–	1,355	296	–	–	1,651
Northern	0.47	–	1,268	–	5,945	7,213
Northwestern	–	–	315	102	9,624	10,041
Southern	0.42	1,024	3,649	–	–	4,673
Western	–	2,679	176	5,190	209	8,254
Zambia	56.9	8,391	11,725	5,292	24,560	49,968
FGS (million m³) (E)	9.21	410	558	251	1,720	2,939.50
CFGS (million m³) (E)	7.37	51	69	31	213	364.50
MAI (m³/ha/a) (E)	5.50	1.00	1.54	0.92	2.00	1.36
AAC (million m³) (E)	0.28	1.01	2.17	0.58	5.89	9.65
Production (million m³) (E)	0.65	–	–	–	–	11.32

Note: (E) indicates estimates, AEZ is the agro-ecological zone, FSG is the forest growing stock, CFGS is the commercial forest growing stock and MAI is the mean annual increment.

Table 1.5 Forest Growing Stock and Production Trends from Natural Forests in Zambia (2001–2010)

Description	U/M	2001	2003	2005	2007	2010
FGS	million m^3	1,461	1,431	1,400	2,940	2,910
– Forests	million m^3	1,362.00	1,335.00	1,307.00	2,785.00	2,755.00
– Woodlands	million m^3	99.00	96.00	93.00	58.00	58.00
– Other natural forests	million m^3	–	–	–	97.00	97.00
CFGS	million m^3	175.32	171.72	168.00	365.80	365.80
– Area	million ha	5.99	5.99	5.99	5.99	5.99
– MAI	m^3/ha/a	1.61	1.61	1.61	1.61	1.61
– AAC	million m^3/a	9.65	9.65	9.65	9.65	9.65
Total Production	million m^3	9.09	9.65	10.22	10.80	13.69
– Fuelwood	million m^3	2.47	2.71	2.97	3.22	3.64
– Charcoal	million m^3	5.48	5.50	5.50	5.50	7.97
– Saw logs	million m^3	0.23	0.51	0.83	1.16	1.16
– Poles	million m^3	0.92	0.92	0.92	0.92	0.92

Note: FSG is the forest growing stock, CFGS is the commercial forest growing stock, MAI is the mean annual increment and AAC is the annual allowable cut.

forest plantations accounting for 52,000 ha. In addition, about 4,000 ha come from local supply forest plantations located in other Provinces and ranging from 240 ha in Central Province to 2,840 ha in Eastern Province. Total FGS from plantations is estimated at 9.21 million m^3 out of which 7.37 million m^3 are considered suitable for commercial utilization. The total round-wood production is 11 and 14 million m^3 per annum from the total forest area of 49,968 million ha (Table 1.4).

Production Trends from Natural Forests
According to Ratnasingam and Ng'andwe (2012), over 80% of the round-wood harvested from natural forests is used as woodfuel (charcoal and fuel-wood). For example, from the estimated production of 13.69 million m^3 in 2010, 11.6 million m^3 was used for woodfuel production (Table 1.5).

Production Trends from Forest Plantations
Forest plantations in Zambia were established from 1965 in all Provinces of Zambia with the Copperbelt Province accounting for over 80% by area (Ng'andwe, 2012a). Industrial and local supply plantations are estimated at 55,000 ha with private plantations which were leased in 2002 covering only 2,000 ha. Zambia Forestry and Forest Industries

Table 1.6 Forest Growing Stock and Annual Allowable Cut in Forest Plantations, 2010				
	Species	Area 1,000 ha	FSG (million m³)	Forest Stocking (m³/ha)
Hardwoods	Eucalyptus grandis	11 000	1.70	160.00
	Eucalyptus cloeziana		0.50	
	Gmelina arborea		0.04	
Softwoods	Pinus kesiya	46,900	4.74	180.00
	Pinus orcarpa		2.20	
	Pinus michocana		0.03	
	Total	56,900	9.21	170.00
Note: FSG is the forest growing stock.				

Table 1.7 Forest Growing Stock and Production Trends from Forest Plantation in Zambia (2001–2010)						
Description	U/M	2001	2003	2005	2007	2010
FGS	million m³	18.406	15.673	14.173	13.786	9.211
CFGS	million m³	14.500	13.250	11.750	11.856	7.369
MAI	m³/ha/a	12.000	11.000	10.500	9.000	5.500
AAC	million m³/a	0.614	0.563	0.537	0.461	0.281
Total Production	million m³	0.199	0.335	0.522	0.584	0.648
— Other logs	million m³	0.063	0.090	0.195	0.298	0.313
— Saw logs	million m³	0.118	0.215	0.291	0.249	0.293
— Chiplogs	million m³	0.007	0.012	0.014	0.016	0.018
— Peeler logs	million m³	0.004	0.010	0.010	0.009	0.011
— Props, billets	million m³	0.003	0.003	0.006	0.006	0.007
— Poles	million m³	0.003	0.003	0.004	0.004	0.004
— Firewood	million m³	0.001	0.002	0.001	0.001	0.002
Note: FGS is the forest growing stock, CFGS is the commercial forest growing stock, MAI is the mean annual increment and AAC is the annual allowable cut.						

Corporation Limited (ZAFFICO) has the mandate to manage forest plantations in Zambia. The FGS from Zambia's forest plantations has declined from 18.4 million m³ in 1990 to 9.2 million m³ in 2010 (Table 1.6).

As wood industries developed, there has been a corresponding upward round-wood production trend from 0.199 million m³ in 2001 to 0.648 million m³ in 2010 against the decreasing annual allowable cut from 0.614 million m³ to 0.281 million m³ per annum during the same period (Table 1.7). The decreasing annual allowable cut is attributed to lack of plantation expansion and replanting and is likely to result in a shortage of round-wood supply to industry in the future.

Over the years, there has been a steady increase in the actual supplies of plantation round-wood from 140,000 m^3 in 2001 to approximately 500,000 m^3 in 2011 (ZAFFICO, 2007). On the other hand, the demand has been 30–50% above the actual supply from 2005 onwards (Nshingo, 2007). With unsustainable management practices and skewed age class structure, the forest plantation growing stock has been depleted over the years, from 18.4 million m^3 in 2001 to 9.2 million m^3 in 2010 (Ng'andwe, 2011, 2012a) (Table 1.7). In addition, the decline in total plantation growing stock and annual allowable cut may largely be attributed to inadequate replanting, reduced tree growth owing to senescence and persistent annual over harvesting trends due to the ever increasing national and regional timber demand.

ZAFFICO has over the years tried to put measures in place in order to address the declining wood resources from forest plantations (ZAFFICO, 2004, 2007, 2013, 2014). The 2014–2018 Forest Plantation Management Plan was developed by the Corporation for its 2014–2018 Strategic Plan using the results from the inventory, socio–economic, timber demand–supply and infrastructure surveys. The 2014–2018 Forest Plantation Management Plan has been geared towards implementing strategies that will enhance annual replanting and expansion of forest plantations to other Provinces in order to ensure sustainable national round-wood supply in the short, medium and long term. The main programs in ZAFFICO's strategic Plan include sustained replanting and forest plantation expansion, wood harvesting and yield regulation, resource mobilization and administration.

Round-Wood Production Statistics

Production of industrial round-wood from forests designated for timber production under the current forest classification system (Ratnasingam and Ng'andwe, 2012) is estimated at 1.813 million m^3 (Ng'andwe et al., 2012). Industrial round-wood production has increased from 427,000 m^3 in 2001 to 1,813,000 m^3 in 2010 due to the increase in demand for timber in construction, renovation and building sectors of the national economy (Table 1.8) (ZAFFICO, 2007).

The source of industrial round-wood production from 1980 to 2010 has been natural forests and plantations which accounted for an average of 71% and 29%, respectively (Figure 1.2).

Table 1.8 Summary Statistics of Industrial Round-Wood Production in Zambia (2001–2010)

Year	Hardwoods (1,000 m³)	Softwoods (1,000 m³)	Total Production (1,000 m³)
2001	227	199	427
2005	827	522	1,349
2010	1,155E	658	1,813

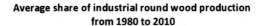

Average share of industrial round wood production from 1980 to 2010

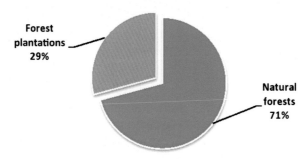

Forest plantations 29%

Natural forests 71%

Figure 1.2 Share of industrial round-wood production in Zambia 1980–2010.

Table 1.9 Summary Statistics of Industrial Round-Wood Production (2001–2010)

Year	Woodfuel (m³)			Industrial Round-Wood (m³)			Total (m³)
	Fuel-Wood	Charcoal	Subtotal	Hardwood	Softwood	Subtotal	
1980	1,943	3,634	5,578	450	12	462	6,040
1985	1,482	4,676	6,158	488	12	500	6,658
1990	1,583	5,174	6,757	563	113	676	7,433
1995	1,782	5,830	7,611	650	300	950	8,561
2000	2,352	5,830	8,182	528	339	867	9,049
2005	2,968	5,830	8,798	827	522	1,349	10,147
2010	3,644	7,969	11,613	1,155	658	1,813	13,426

Round-wood production from natural forests, woodlands and plantations has increased since 1980 with woodfuel accounting for about 89% and industrial round-wood for only 11% of total wood production in Zambia (Ng'andwe et al., 2012). There has been an increase in woodfuel and industrial round-wood production since 1980 indicating the growth in this sector (Table 1.9).

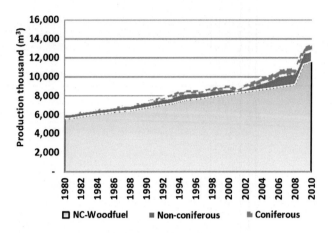

Figure 1.3 Round-wood production trends in Zambia 1980–2010.

Historically, charcoal production increased from 649 MT in 1980 to 1,041 MT in 1994 and the statistics remained constant since then (Table 1.9). This indeed raises a lot of questions as to whether the data on charcoal production is up-to-date (Mwitwa and Makano, 2012; Kewin, 2009; Ng'andwe and Ncube, 2012). For example, charcoal consumption in Eastern, Copperbelt and Lusaka Provinces averages 1,423 MT according to Mwitwa and Makano (2012), while earlier Kewin (2009) estimated an average of 1,392 MT for the whole country. In modeling carbon dioxide emission reduction potential through the use of improved charcoal cook stoves in Zambia, Ng'andwe and Ncube (2012) estimated an average of 2,511 MT for using traditional charcoal cook stoves. These differences in statistics clearly show the difficulty in obtaining accurate statistics that reflect the situation on the ground. On the other hand, the contribution of the forest products industry to the socio-economy of the country is significant as shown by the increasing production trend (Figure 1.3). The round-wood production for woodfuel increased from about 5 million m^3 in 1980 to over 11 million m^3 in 2010 from non-coniferous forests (NC). The NC industrial round-wood from natural forests and from the coniferous forests (plantations) has also been increasing over the years (Figure 1.3).

The increasing production trend may also imply that many people are dependent on forests as their source of livelihood apart from industrial processing. In addition, woodfuel is the largest source of cooking energy at household level and is ever increasing since 1980 (Ng'andwe et al., 2012). There has been an increase in industrial round-wood production from both coniferous and NC forests except in 2001–2002 when a decline was experienced. This decrease has been associated with the export ban on logs by the government (Ratnasingam, 2012; Ratnasingam and Ng'andwe, 2012).

EMERGING ISSUES THAT AFFECT THE FORESTRY SECTOR

The importance of forests as a carbon sink and the negative impact of deforestation on the environment have been recognized by the United Nations Framework Convention on Climate Change (UNFCCC). Satellite images generated over a period of 15 years revealed estimates of the national deforestation rates ranging between 250,000 and 300,000 ha per annum (MTENR, 2008b). The range of deforestation estimates is quite large indicating the difficulties encountered by consultants who have attempted to determine the rate of deforestation based on historical data sets and inventory assessments (Kewin, 2009; MTENR, 2008b). However, Kewin (2009) provided a summary of different approaches used in determining the deforestation in Zambia and estimated the total change of carbon stock of 4.4–7.2 million tons due to deforestation driven by a combination of factors.

The high rates of deforestation currently taking place in the country are related to pressure on the natural resources arising from over-exploitation of forest products, destruction by wild fires, pollution and anthropogenic activities (Vinya et al., 2012). The pressure results in considerable vegetation cover loss and degradation, and are driven by (i) unsustainable woodfuel production, (ii) unsustainable timber production by formal and informal sectors, and (iii) unsustainable agricultural methods. Deforestation therefore reduces the potential of forests as sinks for carbon and causes greenhouse gas (GHG) emissions from all forest types. Mwitwa and Makano (2012) estimated for example, that 0.01 ha of undisturbed forest is cleared annually for every ton of charcoal consumed in Eastern Province. Although environmental and territorial impacts of commercial logging are significant, some studies have established that this may not be a predominant factor for deforestation in mixed forests such as the Miombo woodlands, compared to pure stand forests with one or two commercial species such as *B. plurijuga* and *C. mopane* (Joanne, 1993; Vinya et al., 2012; Mwitwa and Makano, 2012).

REDUCING EMISSIONS FROM DEFORESTATION AND FOREST DEGRADATION

Drivers of deforestation and forest degradation are a result of a complex set of underlying causes that are primarily caused by past and current development processes. These underlying causes cut across numerous sectors (e.g., energy, forestry, agriculture, and water) (Vinya et al., 2012; Mwitwa and Makano, 2012). In order to address them and thereby to facilitate the realization of REDD+, the entire mode of development and the supply and demand chain of wood and NWFPs will need to be altered. According to the state of the world forest report (FAO, 2011), a group of countries moved a motion aimed at Reducing, Deforestation, Developing (RED) countries. This was later expanded to include "reducing emissions from deforestation and forest degradation in developing countries (REDD)" and "conservation, sustainable management of forests and enhancement of forest carbon stocks in developing countries (REDD+)." The scope of REDD+ goes beyond deforestation and forest degradation to include the maintenance and enhancement of forest carbon stocks. It specifically includes:

- reducing emissions from deforestation;
- reducing emissions from forest degradation;
- sustainable management of forests;
- conservation of forest carbon stocks; and
- enhancement of forest carbon stocks.

REDD+ is designed not only to enable developing countries to contribute to a reduction in emissions under future arrangements to the UNFCCC, but also to strengthen SFM at local and national levels. The timber processing industry of Zambia faces challenges associated with the climate change, but at the same time creates new opportunities for development. International efforts over the past two decades to build a common understanding as well as a policy framework and a range of tools for sustainable forest management (SFM) provide a sound basis for policymakers and forest managers to address the climate change effectively. Zambia has been participating in the REDD+ programs since 2010 through pilot projects (GRZ, 2010c).

The "REDD Implementation Framework" in the "REDD+ Strategy" includes the design of a transparent and equitable benefit sharing system which takes into cognizance: (i) Stakeholder engagement to build ownership, trust and legitimacy, (ii) timing and adequacy of benefits to cover opportunity costs, (iii) safeguards against corruption in reporting, auditing, and monitoring flow of benefit, (iv) transparency in the transactions of benefit, and (v) flexibility of agreements to facilitate adaptation and clear dispute settlement mechanism. The challenge for this proposal is to develop a system that provides benefits that are effective in reducing GHG emissions from deforestation and forest degradation.

CARBON MANAGEMENT AND TRADE

In the rush to reveal the prospects of carbon management in Zambia's forests, alternative and complementary opportunities in the production of wood and NWFPs cannot be overlooked (Ng'andwe, 2012b). Forests in Zambia cover approximately 60% of the total land area and are natural and potentially effective in reducing GHGs emissions. With chances for a new comprehensive global climate treaty that emerged after Copenhagen and Cancun, consideration of both the UN Clean Development Mechanism (CDM) and Voluntary Carbon Markets (VCM) producing carbon offsets remains prudent even in Zambia. The complexities and time-lags evident in the CDM project registration process and its coverage of only afforestation and reforestation (A/R) projects in the forestry sector, has seen considerable attention being diverted to the voluntary carbon markets. This is where the project-based avoided deforestation activity is helping lead development of the REDD+ program.

Zambia is one of the nine pilot countries for the UN-REDD program and as the country develops a national REDD strategy, it is possible that many of the forests could be managed as carbon forests and therefore there is a need to have clear guidelines for benefits sharing (BS) that include communities. Experiences from existing BS mechanisms in forest conservation and management systems, such as Integrated Conservation and Development Projects, Payment for Environmental Services, Community Forestry Management, and Production Forestry will provide a basis for assessing their functionality and whether modified or new mechanisms are needed under CDM and voluntary carbon markets.

According to the State of the World Forests Report (FAO, 2009) payment for carbon sequestration to mitigate climate change is one of the fastest-growing environmental markets. Under the Kyoto Protocol, the CDM may offset a certain part of their emissions through investment in carbon sequestration in industrialized Annex I countries or substitution projects in non-Annex I of the Protocol (i.e., developing) and thus acquire tradable certified emission reductions. Under joint implementation, Annex I countries of the Protocol may jointly execute carbon sequestration or substitution projects, thus emission trading permits the marketing of certified emission reductions.

In addition, the appeal of afforestation and reforestation as a climate change mitigation strategy is considerable in Zambia but forest-based carbon offset projects face several challenges including setting baselines, permanence, and leakage and monitoring constraints. The problems are particularly severe in certain parts of Zambia with high deforestation rates coupled with policy and institutional constraints. These issues have hindered a more prominent role for forests in climate change mitigation under the CDM. Identification of all relevant stakeholders (government, formal and informal forest users, private sector entities, individuals, local communities) early in the process ensures their adequate participation in any benefit-sharing scheme and avoidance of leakage. Enhanced stakeholder participation in the management of carbon resources will result in the provision of greater benefits to local communities as agreements and laws become more transparent.

FOREST AND TIMBER CERTIFICATION

Forest and timber certification is considered to be a market based instrument for improving forest management. Today, certification, as a market based instrument, has become popular in government, non-government, trade, and industry worldwide. Forest certification is a separate initiative following the United Nations Conference on Environment and Development in Rio de Janeiro in 1992 which recognized that problems of poverty and food security were linked to deforestation and indebtedness of developing countries. Since 1992 there have been a number of intergovernmental approaches that indirectly provided the natural starting point for the development of certification standards. On the other hand, SFM is considered as a key objective in forestry policy formulation. However, Forest certification started as a separate initiative by some Environmental Non Governmental Organizations (ENGOs) and interest groups in 1993, by-passing the intergovernmental approaches. The same year the Forest Stewardship Council (FSC) was established and from 2001 onwards other timber certification schemes, eco-labels and claims proliferated. The FSC scheme is more prominent and international, and has achieved worldwide recognition (Ng'andwe, 2003; Njovu, 2004). On the other hand, the Program for Endorsement of Certification is also gaining international recognition.

While Forest certification has been identified as the key tool to bring about SFM others disagree saying that it could simply lead to trade diversion (Cadman, 2002). Proponents argue that unresolved factors regarding nature are influencing people towards a more stewardship type of relationship and the areas of certified forests are increasing globally. For example, FSC certified forest has increased from 140 million ha in 2004 to over 180 million ha by the end of 2013. In Africa the FSC scheme is prominent, with South Africa having the largest share of certified plantations of over 1 million ha. The Republic of Congo has the largest share of tropical certified forests with over 2.4 million ha, followed by Gabon with 1.8 million. On the other hand, Brazil dominates with over 7.2 million ha of certified forests.

Zambia was the first country to have parts of its forests certified in Africa in 1998 by the FSC for 1 million ha under the Muzama Craft community based hardwood forest but lost the status in 2000 for non-compliance. Certification in Zambia developed not as a government process but as a donor driven initiative to export honey and bee products. The lessons learned motivated other donors to fund forest certification under the FSC Scheme of the community based hardwood project of the Muzama Crafts Limited in 1998. The first local private sector initiative to certify in 2003 arose from the increase in the demand for certified wood products on the US market, the need to stay in the market and higher product price expectations. Today, many companies desire to be certified but face challenges and dilemmas of an old forest policy, the cost of bringing their forest management practices to the FSC standard, the cost of certification, and lower demand for certified plantation based soft-wood products. Nevertheless, there has been an improvement in the social image and performance of the companies who have tried certification. This development has generated a lot of interest for other producers to get certified and stakeholders now view certification as the only tool that can be used to sustainably manage and utilize 50,000 ha of plantation resources and 49 million ha of natural forests whose potential supply is declining rapidly due to over cutting.

The driver of forest certification in Zambia largely has been for market access to the eco-sensitive markets of North America and

Europe. Most African countries who have been exporting timber to Europe face increasing challenges of certifying their forests with credible systems recognized by the buyers in order to continue exports. Forest certification has since grown, the supply of certified hardwood and plantation raw material and products is assured but demand for certified products needs to be created. The way forward for Zambia is to create awareness in government, institutions and the private sector by collaborating with ENGOs through national initiative strategies.

FOREST LAW, ENFORCEMENT, GOVERNANCE AND TRADE

In recent years there has been increased awareness of the economic, social and environmental contributions of forests to livelihoods. Until recently the forest sector has been considered as a low priority sector in Zambia. Nevertheless, the forestry sector status is changing with renewed and increased attention brought about by emerging issues such as climate change, carbon trade, and renewable energy issues and is now being ranked among the priority sectors in Zambia. At the same time Forest Law and Enforcement, Governance and Trade (FLEGT) attempts to curb illegal logging. The sixth national development plan (GRZ, 2011) and the vision 2030 (GRZ, 2011) aspire to have a forest sector that takes advantage of the opportunities available including those related to policy shifts, legislation, and private sector involvement. The forest and logging as well as downstream processing activities contribute about 5.2% to the Gross Domestic Product and therefore Zambia aspires to increase the contribution of the forestry sector to the national economy and employment creation.

In Zambia and other countries in Africa, trade in timber products is characterized by unprocessed to semi-processed products, primarily roundwood and sawn boards, which indicate that the full value of forest resources is not captured. This scenario presents an opportunity for investing in primary processing and value-adding of wood products amid major challenges such as illegal logging. The rampant exploitation and underutilization of harvested indigenous timber compelled the Government of Zambia to temporarily suspend all timber licenses in 2012 "to protect the depleting forests around the country" (Ratnasingam and Ng'andwe, 2012).

Illegal logging poses a major challenge for the establishment and maintenance of efficient markets and sustainable logging practices in a global economy that increasingly demands assurances of legal and sustainable production of wood and wood products. In 2003, the European Commission adopted the FLEGT Support Program Action Plan whose ultimate goal is to encourage sustainable management of forests. To these ends, ensuring the legality of forest operations is a vital first step. The Plan focuses on governance reforms and capacity building to ensure that timber exported to the European Union comes only from legal sources. According to the FEGT action plan Illegal behavior in the logging sector results in lost government revenue, missed opportunities for industrial development, and increased environmental damage and social problems. The FLEGT initiative if pursued by the Government can reduce the rampant exploitation of the commercial and lesser known timber species under threat.

CONCLUSION

The forestry sector of Zambia includes forest and logging and the downstream processing activities involving wood and NWFPs. The ZFAP of 1997 (ZFAP, 1997), Forest Policy of 1998, and Forest Act of 1999 (GRZ, 1999) provide for wood industry development and community participation in forest management and utilization. The NEAP of 1994 led to the development of the Forest Policy of 1998 which is aimed at changing the national institutional and legal framework for forest management and administration so that the forestry sector could be administered by a forest commission.

The contribution of the forestry sector to the national economy is significant given the role the sector plays in supporting livelihood at various levels. The FAO agro-ecological zoning as a system that defines specific ecosystems based on soil, landform, and climatic characteristics for tree growth and land management systems (FAO, 2002) has been used in Zambia to identify different zones in relation to forest and woodland productivity. In the ILUA of 2005–2008, three main zones were identified and the respective potential productivity of forests determined based on the inventories for each Province (MTENR, 2008b; GRZ, 2010b). In order to link different forest classes and AEZ to production potential it is important that the annual allowable cut is determined during inventories for each forest class. It is also useful for investors to know the CFGS of commercial species and lesser used species from both natural forests and forest plantations by forest type and ownership (e.g., how much customary ownership, protected forests). The forestry sector faces opportunities for enhancing marketing of wood products and NWFPS, particularly the incentives from Government aimed at establishing a profitable environment for increased domestic industrial growth, export promotion, the development of market-oriented production management, and private sector development. On the other hand, the timber processing industry of Zambia faces challenges associated with climate change which in another way presents new opportunities for policymakers and forest managers to address the climate change issues effectively, access global markets, and manage their resources sustainably.

REFERENCES

Brown, C., 1999. The Outlook for Future Wood Supply from Forest Plantations. Global Forest Products Outlook Study Working Paper GFPOS/WP/03. Food and Agriculture Organization of the United Nations, Rome.

Cadman, T., 2002. Sustainable forest management and timber certification. Will timber certification deliver sustainable forest management? <http://www.nfn.org.au/cat01.htm> (accessed 12.12.14.).

Campbell, B., 1996. The Miombo in Transition: Woodlands and Welfare in Africa. CIFOR, Bogor.

Chidumayo, E.N., 1996a. Handbook of Miombo Ecology and Management. Stockholm Environment Institute, Stockholm, Sweden.

Chidumayo, E.N. 1996b. Land Use Planning. The Zambia Forestry Action Programme (ZFAP) Secretariat. Lusaka, Zambia.

Chidumayo, E.N., 2012. Classification of Zambian Forests under Report Prepared for the Integrated Land Use Assessment (ILUA) Phase II Project. Lusaka, Zambia: Makeni Savanna Research Project.

Chileshe, A., 2001. Forestry outlook studies in Africa (FOSA)—Ministry of Natural Resources and Tourism—Zambia. Available online: <http://www.fao.org/forestry/FON/FONS/outlook/Africa/AFRhome>.

Chisanga, E.C., 2005. Establishment and status of forest plantations in Zambia. In: Ng'andwe, P. (Ed.), First National Symposium on the Timber Industry in Zambia. Mulungushi International Conference Center: Mission Press, Lusaka, Zambia, 29–30 September.

FAO, 2002. Food and Agriculture Organisation of the United Nations Global Agro Eocological Zoning System. FAO, Rome, Italy.

FAO, 2009. Food and Agriculture Organisation of the United Nations State of the World Forests Report. FAO, Rome, Italy.

FAO, 2010. Food and Agriculture Organisation of the United Nations. Global Forest Resource Assessment. FAO, Rome, Italy.

FAO, 2011. Food and Agriculture Organisation of the United Nations State of the World Forest Report. FAO, Rome, Italy.

GRZ, 1965. In: Forestry (Ed.), Forest Policy of 1965. Government Printers, Lusaka, Zambia.

GRZ, 1973. In: Forestry (Ed.), Forest Act No. 39 of 1973. Government Printers, Lusaka, Zambia.

GRZ, 1995. In: ZDA (Ed.), Zambia Development Agency Investment (Amendment) Act of 1993. Government Printers, Lusaka, Zambia.

GRZ, 1997. In: MENR (Ed.), Zambia Forestry Action Plan. Draft Policy. Ministry of Environment and Natural Resources—Lusaka, Zambia. Government Printers, Lusaka, Zambia.

GRZ, 1998a. Government of the Republic of Zambia. Draft Lands Policy of 1998. In: Lands (Ed.), Lusaka, Zambia.

GRZ, 1998b. In: Forestry (Ed.), Forest Policy of 1998. Government Printers, Lusaka, Zambia.

GRZ, 1999. In: Forestry (Ed.), Draft Forest Act of 1999. Government Printers, Lusaka, Zambia.

GRZ, 2006. In: GRZ (Ed.), Government of the Republic of Zambia Vision 2030. Government Printers, Lusaka, Zambia.

GRZ, 2010a. In: Forest (Ed.), Government of the Republic of Zambia. Draft National Forest Policy of 2010. Government Printers, Lusaka, Zambia.

GRZ, 2010b. Government of the Republic of Zambia, Integrated Land Use Assessment 2011–2013. In: MTENR (Ed.), Lusaka, Zambia.

GRZ, 2010c. Government of the Republic of Zambia. UN Collaborative Programme on Reducing Emissions from Deforestation and Forest Degradation in Developing Countries. National Joint Programme Document—Zambia Quick Start Initiative. In: MTENR (Ed.), Lusaka, Zambia.

GRZ, 2011. In: MOFNP (Ed.), Zambia Sixth National Developmemt Plan_Final_Draft. Government Printers, Lusaka, Zambia.

IDLO, 2011. International Development Law Organisation. Legal preparedness for REDD in Zambia. Rome, Italy.

Joanne, C.B., 1993. Timber Production, Timber Trade and Tropical Deforestation, Biodiversity, Ecology, Economics, Policy (May 1993), vol. 22. Ambio, pp. 136–143.

Kewin, B.F.K., 2009. Carbon stock assessment and modelling. In: Zambia-A UN REDD Programme Study. UNDP, FAO and UNEP, Lusaka, Zambia.

Meneses-Tovar, C.L., 2011. NDVI as an indicator of degradation. Unasylva 62 (238), 2011/2.

MTENR, 2008a. Ministry of Environment and Natural Resources Management and Mainstreaming Programme. In: FD (Ed.), Lusaka, Zambia.

MTENR, 2008b. Integrated Land Use Assessment 2005–2008. Forestry Department, Ministry of Tourism, Environment and Natural Resources, Lusaka, Zambia.

Mwitwa, J., Makano, A., 2012. Charcoal Demand, Production and Supply in the Eastern and Lusaka Provinces. Mission Press, Lusaka, Zambia, Ndola.

Ng'andwe, P., 2011. Round Wood Supply and Demand Trends in Zambia and the SADC Region. Consultant Report Submitted to Zambia Forestry and Forest Industries Corporation, Ndola, Zambia. Kitwe, Zambia: Copperbelt University.

Ng'andwe, P., 2012a. Forest Classification, Zones and Classes—Basis for Industrial Processing. Submitted to the Ministry of Lands, Natural Resources and Environmental Protection and FAO, Lusaka, Zambia.

Ng'andwe, P., 2012b. Forest Products Industries Development—A Review of Wood and Wood Products in Zambia. Submitted to the Ministry of Lands, Natural Resources and Environmental Protection and FAO, Lusaka, Zambia.

Ng'andwe, P., Ncube, E., 2012. Modeling carbon dioxide emissions reduction through the use of improved cook stoves: a case for Pulumusa, portable clay and fixed mud stoves in Zambia. UNZAJST 15 (2), 5–15.

Ng'andwe, P., Simbangala, L., Kabibwa, N., Mutemwa, J., 2012. Zambia biennial compendium of forestry sector statistics 1980–2010. Compendium, Ndola, Zambia.

Ng'andwe, P., 2003. Timber Certification, Optimization and Value Added Wood Products—A Case Study for Zambia Master of Science Forest Industries Technology. University of Wales, United Kingdom.

Njovu, F., 2004. Forest certification in Zambia. Paper Presented at the Symposium on Forest Certification in Transition Countries: Social, Economic and Ecological. Yale School of Forestry and Environmental Studies, New Haven, CT, USA.

Nshingo, C., 2007. Long Term Trends in the Timber Sector, Emerging Opportunities and Challenges for Sustainable Plantation Forest Management in Zambia. MBA, Copperbelt University, Kitwe, Zambia.

Ratnasingam, J., 2012. The Status of the Wood Products Sector in Southern Africa. IFRG Report No. 14, Singapore.

Ratnasingam, J., Ng'andwe, P., 2012. Forest Industries Opportunity Study—Synthesis report submitted to the Forestry Department Integrated Land Use Assessment II and the Food and Agriculture Organisation (FAO) of the United Nations.

UN, 2006a. The International Industry Classification of Economic Activities ISIC Revision 4 Statistical Papers. Agriculture, Forestry and Fishing. United Nations Department of Economic and Social Affairs Statistical Office, New York, NY, USA.

UN, 2006b. Standard International Trade Classification (SITC) Rev 4. No. E.06.XVII. 10. Statistical Papers Series. United Nations Department of Economic and Social Affairs Statistical Office, New York, NY.

Vinya, R., Syampungani, S., Kasumu, E.C., Monde, C., Kasubika, R., 2012. Preliminary study on the drivers of deforestation and potential for REDD + in Zambia. A Consultancy Report Prepared for Forestry Department and FAO under the National UN-REDD+ Programme Ministry of Lands & Natural Resources, Lusaka, Zambia.

Whiteman, A., Brown, C., 1999. The potential role of plantations in meeting future demand for industrial round wood. Int. For. Rev. 1 (13), 143–152.

ZAFFICO, 2004. Zambia Forestry and Forest Industries Corporation Forest Plantation Management Plan. In: Zimba, S.C., Ng'andwe, P., Njovu, F., (Eds.), Management Plan 2004–2008. Ndola, Zambia.

ZAFFICO, 2007. Zambia Forestry and Forest Industries Corporation Management Plan. In: Ng'andwe, P., Nshingo, C., Chisanga, E., Njovu, F., (Eds.), Management Plan 2008–2013. Ndola, Zambia.

ZAFFICO, 2013. Zambia Forestry and Forest Industries Corporation Strategic Plan 2014–2018. Ndola, Zambia.

ZAFFICO, 2014. Zambia forestry and forest industries corporation management plan. In: Ng'andwe, P., Nshingo, C., Chisanga, E., Njovu, F., Kasubika, R. (Eds.), Management Plan 2014–2018. School of Natural Resources, Copperbelt University, Kitwe, Zambia.

ZFAP, 1997. Zambia Forestry Action Plan 1997–2015, Forestry Department. In: MENR (Ed.), Lusaka, Zambia.

CHAPTER 2

Wood and Wood Products, Markets and Trade

Phillimon Ng'andwe[a], Jegatheswaran Ratnasingam[b], Jacob Mwitwa[c], and James C. Tembo[a]

[a]Department of Biomaterials Science and Technology, School of Natural Resources, Copperbelt University, Kitwe, Zambia; [b]Faculty of Forestry, University Putra Malaysia, Serdang, Selangor, Malaysia; [c]Department of Plant and Environmental Sciences, School of Natural Resources, Copperbelt University, Kitwe, Zambia

INTRODUCTION

Wood is a renewable and highly versatile natural resource that fuels a wide range of labor-intensive processing activities providing employment to almost 1.1 million people (Ratnasingam and Ng'andwe, 2012). The processing capacity of Zambia for round-wood is estimated at 1,500,000 m^3 per annum of wood and various wood products (Ratnasingam and Ng'andwe, 2012). However, the subsector is predominantly characterized by inefficiency, poor quality, and low productivity due to insufficient technical skills and know-how. Zambia's wood industry suffers from outdated technology and poor product quality that severely limits access to markets in the region.

The hardwood-based mills were specifically located in Western (Mulobezi sawmills) and Copperbelt Provinces (Minga mining timbers) during the 1960s with a combined installed sawn timber production capacity of about 500,000 m^3 per annum. Considering the country's huge forest wood resources, there is an opportunity to significantly increase this capacity through selective investment into greater value-added manufacturing and marketing activities also focusing on the lesser-known wood species.

The opportunities in the forestry sector markets are increasing in line with high demand in domestic and regional markets. The government's policy to stimulate private sector development initiatives and the provision of a good investment environment have created conducive market incentives that would attract investment opportunities in the wood industry. This Chapter presents a synthesis of information relating to the wood

Forest Policy, Economics, and Markets in Zambia. DOI: http://dx.doi.org/10.1016/B978-0-12-804090-4.00002-1

industry development, current production, and trade potential. Furthermore, the chapter highlights challenges that constrain the development of wood industry and how policy can bring about change.

WOOD INDUSTRY DEVELOPMENT

The mining industry has been the major driver of the wood industry development in Zambia. Hardwood industry was developed first to support the mining needs for wood and also for railway sleepers. The first and second national forest inventories of 1952 and 1965 provided data for the planning of wood exploitation and forest protection at district and national levels in Zambia (MTENR, 2008). The Forest Act No. 39 of 1973, as a legal instrument, provided for a legal framework for developing wood processing industries (WPIs) in Zambia (GRZ, 1965, 1973, 1998). With the legal framework in place Zambia embarked on the country-wide establishment of exotic tree plantations based on the 1965 forest inventory (Chisanga, 2005).

With the establishment of forest plantations, the softwood products industry developed in the Copperbelt Province around the 1990s targeting the mining industry. The commercial forest growing stock from industrial plantations was about 600,000 m^3 per annum by 2002, with an additional 100,000 m^3 per annum from Local Supply Plantations established in provincial centers (Ng'andwe et al., 2012). Processing units such as pole treatment plants; sawmilling and kiln-drying facilities; and plywood, veneer, and blockboard manufacturing factories were developed around the country between 1992 and 1995. These processing units were mainly government-owned companies, established as parastatals.

During 1980–1992, the timber industry was dominated by state-owned enterprises and by 1990s most of these parastatals were performing badly. In about 1993 the Zambian government commenced major economic reforms focusing on the privatization of state-owned enterprises and liberalizing the economy. Following this macro-economic shift, the largest government-owned timber processing plants of the Zambia Forestry and Forest Industries Corporation (ZAFFICO) and Zambia Steel and Building Supplies (ZSBS) were privatized. According to a government report (GRZ, 2006a) privatization and structural adjustment programs resulted in the deterioration of the productivity in the manufacturing sector. Many employed persons in the formal sector were retrenched.

The retrenched personnel, as a result of the Structural Adjustment Program of 1993–2000, started developing private small-scale forest-based sawmilling enterprises throughout the country as a means of livelihood (Njovu, 2011). Since 2001, there has been a modest small-scale private investment in sawmilling, particleboard, plywood, and pole treatment technologies. In the subsequent years, there has been a mushrooming of informal wood processing segments particularly in the Copperbelt and Lusaka Provinces providing additional employment. Since then, demand for industrial round-wood increased from 140,000 m^3 in 2001 to over 650,000 m^3 by 2010 (Ng'andwe, 2011a; Ng'andwe et al., 2011). On the other hand, technological development lagged behind as many sawmills continue to use obsolete machines characterized with low recoveries and low quality products. The lack of meaningful investment resulted in a policy development that would catalyze investment in the timber industry driven by the private sector.

In 2004 the Government of Zambia initiated the Forestry Department Credit Facility (FDCF) to address issues of lack of investment in the timber industry (Masinja, 2005). In most cases the lack of investment has been attributed to high interest rates, high collateral requirements and reluctance by Banks to lend money to entrepreneurs. The FDCF facility catalyzed the formation of over 500 enterprises engaged in sawmilling as a business in Zambia. The major downstream formal timber products processing units created by the fund were in sawmilling, wood furniture, and Woodcrafts. Today, sawmilling is by far the most dominant economic activity in Zambia's wood industry with over 500 actors in the primary timber processing followed by the wood-based panels (WBPs) industry and wood furniture. The informal wood processing sector on the other hand, has over 1 million actors but has not benefited from the FDCF arrangement.

WOOD INDUSTRY CLASSIFICATION

According to the system of national accounts, wood and wood products processing is classified under manufacturing (CSO, 2008a) and not forestry. However, the forestry sector as used in this book includes downstream processing activities based on the system of classification developed by the International Standard Industry Classification (ISIC) of economic activities of the United Nations (UN, 2006.). This classification was domesticated in Zambia in 2008 for the purpose of accounting for

Figure 2.1 The forest plantation timber production value chain. (a) Mature pine forest plantation for harvesting, (b) Bell equipment sawlog extraction and loading system, and (c) timber shed and office.

the total forestry sector contribution to the national economy and poverty reduction. For example, the forest value chain starts from the forest and logging to the manufactured wood product (Figure 2.1). The domesticated classification has no boundary between forestry and downstream processing activities for the purpose of accounting for the sector's aggregated contribution to national economy (Ng'andwe et al., 2008).

The timber industry provides employment in management, logging, processing, and downstream construction industry, even though the value chain of round-wood products is still not fully developed. There are few big companies including WPIs and Bisonite Plc in Ndola, Copperbelt Forestry Company, Kitwe Wood and Log in Kitwe who have invested in forest equipment and machinery that maximizes the utilization of wood along the value chain.

SAWMILLING INDUSTRY

The sawmilling industry in Zambia has grown from less than 20 mills in 1980 to over 1,000 mills by 2012 (Ratnasingam and Ng'andwe, 2012). In 2008 there were over 100 operators of small-scale sawmills and about 10 medium- to large-scale sawmills (Banda et al., 2008), and by 2012, the number increased to over 900 sawmillers using plantation round-wood and over 100 operators processing hardwood timber (Ng'andwe, 2012). The small-scale sawmills are characterized by the fixed type of push-bench low technology mills and pit-saws supplying low grade timber to traders who sell to the local markets and also export to neighboring countries, such as the DR Congo. Pit sawyers around the country often produce cants, from natural forests, which are sold on the open market such as Buseko in Lusaka for further processing (Figure 2.2).

Figure 2.2 Informal timber processing using pit-sawing method to produce cants which are later sold at Buseko open timber market in Zambia. (a) Pit-sawing technique (b) Timber market, Lusaka, Zambia.

Sawn timber for building construction, renovation and maintenance constitutes the main product from sawmills and is traded downstream for various other end uses such as furniture, paneling, crates, pallets, mine lug boxes, cable drums, etc. Often sawn wood from most sawmills is not kiln-dried (Ratnasingam, 2012a,b; Ratnasingam and Ng'andwe, 2012). An international phytosanitary specification was introduced in Zambia in 2002 requiring timber to be heat-treated or fully dried to the standard of 14% moisture content (Banda et al., 2008), but only about two or three companies such as WPIs and Copperbelt Forest Company (CFC) comply with this voluntary standard. For example, pallets used for export are required to comply with the set standard making it imperative for sawmills to install drying kilns if they are to produce pallets for the export markets. The imposed requirement constitutes a constraint to expand markets by the small-scale operators who continue to experience difficulties in acquiring kiln-drying equipment and technology.

Products from large-scale companies such as WPI, CFC and Bisonite, are kiln-dried and sold with moisture content of 12−15% whereas timber from small-scale producers is sold in wet form. This is mainly because air drying takes time and customers do not demand kiln dried sawn timber amid lack of regulations regarding the use of sawn timber (Ng'andwe, 2005a; Ng'andwe et al., 2006; Ratnasingam, 2012a,b).

WOOD BASED PANEL INDUSTRY

There are two types of mills in Copperbelt Province producing particleboards and one mill producing veneer, plywood, and blockboards, with an estimated combined installed capacity of 50,000 m^3 per annum. The value-added processed products in this industry include:

1. Overlayed particleboard with sliced veneer especially from *Pterocarpus angolensis*.
2. Overlayed particleboard with white and wood imitation foils (beech, mahogany, sapele, etc.).
3. Embossed (decorative) particleboard for ceilings.
4. Shutterboards for housing construction.
5. Overlayed plywood with decorative sliced veneer used as doorskins.

About 95% of value added WBP products are sold in the domestic market and about 5% are exported to neighboring countries.

VALUE-ADDED ENGINEERED WOOD PRODUCTS INDUSTRY

According to Ng'andwe (2005a), the manufacturers of wood products in Zambia are likely to face distinctly different resources in the future as a result of the undergoing depletion of the natural forests and forest plantations resources. The dwindling wood resource will compel producers to start value addition to by-products which include chips, trims, sawdust, shavings, and bark. For example, in value-addition technologies, short pieces and trim blocks are recovered and finger-jointed into high-quality products. Engineered wood products (EWPs) are produced by the rearrangement of wood with the assistance of an adhesive. This is not a new concept as for example, laminated beams have been assembled first with nails, then with adhesives since the 1890s. Trusses, originally assembled with nails, experienced phenomenal growth which started half a century ago with the assistance of steel plates. These products are produced to standards that meet the engineering specifications at CFC in Kitwe and Norzam Gluelam Limited in Ndola. The main value added wood products include: Finger-jointed and glue-laminated beams, edge-glued panels, builder's carpentry, and wooden furniture (Ng'andwe, 2005a; Banda et al., 2008). The application of appropriate wood production technology in the industry is the key to improving the sub-sector recovery in the long run (Box 2.1).

Box 2.1 Modern Value-Addition Technology in Wood Industry

Due to declining supplies of round-wood from plantations and natural forests, any form of growth in the wood industry will have to be based on improved utilization of the existing resource and process diversification towards value-added products. Wasteful processing that produces more forest and mill waste (a) than saleable products will eventually be phased out.

The main by-products of the sawmilling industry such as sawdust, shavings, trims, edging, slabs and bark can be utilized (a). Using appropriate technology, short pieces and trim blocks can be recovered and finger-jointed into high-quality products that meet standards of engineering materials. These products are called non panel value added engineered wood products designed to replace traditional wood products such as panel products for planks and laminated beams for large sections. Engineers make use of design systems by considering use of available resources such as sawmill waste while providing products better suited to market demand. Engineered wood products are assembled according to engineering principles to provide the most efficient match between the forest resource and specific needs of the building industry and demand for high-performance structural products.

Today, appearance wood products are becoming increasingly common. To manufacture these products all types of defects such as knots, wane, rot, soft pith, etc. are eliminated through scanning and optimization techniques (c). The blocks (c) are latter finger jointed into blanks (e). Vertical and horizontal finger jointing machines are used in which the blocks are positioned on edge. Blocks are then clamped together and run past cutters that profiles the blocks (d), then through the glue applicator before being pressed into desired lengths of finger jointed blanks (e). The key value-adding process, after finger jointing is to profile blanks into specific shapes such as mouldings, beams and wall panels. Glue-lamination techniques are used to produce solid wood panels and Glulams, I-beams and builders joinery. Glulams are used for engineered wood construction and optimizes the structural values of a renewable resource. The demand for these products will continue to increase with the surge of the construction activities in Zambia and the continuing desire for modern buildings with modern architectural designs in the region.

In the medium to long term, improved utilization of the resource will be backed up with improved forest management in order to produce short rotation tree crops and lesser known timber species better suited to the needs of the growing industry, and maximization of mill recoveries that enhances value addition.

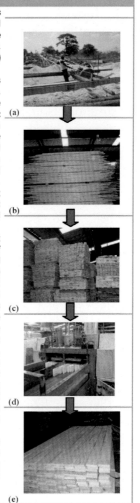

(a)

(b)

(c)

(d)

(e)

WOOD FURNITURE INDUSTRY

The growth of the wood furniture industry has been quite slow since the privatization of government-owned parastatals during the economic reforms and structural adjustment programs of the 1990s. Sawn wood input in the production of wood furniture in Zambia is estimated at 14,523–21,000 m^3 per annum (Puurstjärvi et al., 2005; Ng'andwe et al., 2006). The government-owned companies such as Furniture Corporation of Zambia and ZSBS were the major producers of wooden furniture, but they were later closed down or privatized. Puustjarvi et al. (2005) reported that the private company like Apollo Enterprises and Mukatasha Timbers in Lusaka Province were among the major producers of furniture during 2005.

Currently, most of the quality furniture used in many households and institutions is imported from South Africa while low quality wood furniture is produced under informal processing segments. Wood furniture under this arrangement is made from wet or air dried timber at household level using simple carpentry tools due to limited funds (Campbell, 1996). The informal furniture production is characterized by low skill, limited choice of wood species input and low returns (Campbell, 1996; Banda et al., 2008; Ratnasingam and Ng'andwe, 2012). The wood species commonly used are *P. angolensis*, *Senecio madagascariensis* and *Brachystegia* species and these are often illegally harvested from the communal Miombo woodlands. Several carpentry items are produced including handles for implements, yokes, cooking sticks, mortars, and coffins. However, some informal segments are well organized and use home-made routers and machining tools to produce wood furniture components such as table and chair legs. Some of these processing activities are conducted at wood mass markets such as Buseko in Lusaka, Chisokone in Kitwe and Masala market in Ndola.

On the other hand, large carpentry items such as tables, kitchen units, chairs, doors, windows and finished furniture require massive investment in tools, machines and equipment. The Norzam Gluelam Limited Company produces engineered doors, table tops, windowsills and school desks using a combination of softwood, hardwood, and veneer. The company invested in the latest technologies to enhance the value addition to its production. Such modern technologies include molding machines with four to eight spindles capable of combination planning and profiling in one operation, edge lamination that uses external and internal grade adhesives for bonding and profile wrapping that enhances the appearance of builders joinery and carpentry by a combination of foils, veneer, and vinyl materials. CFC also uses modern technology in value-added production of wooden products such as hardwood doors, door frames, and windows. Mukatasha Enterprises and Apollo Enterprises are also producers of hardwood doors, school furniture, and kitchen units, while ITT in Livingstone produces parquet flooring and wooden furniture.

WOODCRAFTS INDUSTRY

The woodcrafts industry is found in many of the touristic locations in Zambia, particularly in Livingstone, Choma, Mongu, Mumbwa, Serenje, Lusaka Districts and in some parts of Northwestern Province (Puustjarvi et al., 2005). The only known formal woodcrafts firm in the country was the Muzama Crafts Limited which was owned by the pit sawyers, the community and three district councils in Northwestern Province Manginga area. During 1998–2000, Muzama was producing Woodcrafts for export to the United Kingdom and Germany from a certified forest in Northwestern Province. The company stopped producing after the pit sawyers' licenses were not renewed due to non-compliance with the Forest Stewardship Council (FSC) standards. Consequently, Muzama Crafts could not maintain both the forest management and chain of custody certificates due to non-compliance and the forest user right was canceled by the Forestry Department (Ng'andwe, 2005b).

In general, the wood carving and crafts industry is conducted at household level with one to two family members participating and women and children handling the sanding and polishing (Campbell, 1996). The wood raw material used is also often illegally harvested by wood crafters. After Woodcrafts products have been made, they are taken to various marketplaces that are often found by the road side and in most tourist centers. In Zambia, Puustjarvi et al. (2005) estimated sales of over USD 132,000.00 per year to tourists at producer level while in Mpumalanga, in South Africa, Campbell (1996) estimated sales of about USD 100,000.00 per year. The Woodcrafts industry is influenced by many factors including seasonality for tourists, dryness of wood and household subsistence activities mainly agriculture work. The wood crafters also face a combination of constraints including lack of business during given periods, organization skills, lack of capital, limited market knowledge, and depletion of key tree species used in crafting. In many cases such products cannot sell in green markets. For example, green tourism markets are generally based on the principles of sustainable resource management. In such a market products and services are sold in such a way that the resource is not depleted or permanently damaged and can easily be traced from its origin.

EMPLOYMENT IN THE WOOD INDUSTRY

The informal sector contributes significantly to employment in Zambia. According to the industry survey, Ng'andwe et al. (2006) estimated that the number of persons employed in the forest and logging were 1,750 while in manufacturing they were 9,000. In addition, the number of persons employed in wood energy and NWFPs gathering and processing were 152,000 and 888,806 persons, respectively, bringing the total forestry sector employment to 1,051,656 number of persons. In 2008 Zambia's total population was estimated at 12,363,879 out of which only 5,413,518 persons (aged 15 years and above) were in the labor force category while 1,344,931 were inactive (CSO, 2010). The size of the labor force grew by 10% from 4.9 million in 2005 to 5.4 million in 2008 with more males participating (88%) than females (77%).

The country's statistics in the agriculture, forestry and fishing sector indicate that forestry and fishing employment increased from 2,476,000 persons in 1998 to 3,811,922 in 2008 (CSO, 1998, 2004b, 2005, 2008b). In the informal sector the number of persons employed was 90% of the total number of persons employed in Zambia. According to CSO (2008a,b) the informal sector is part of a bigger entity called the Non-Observed Economy (NOE). The NOE corresponds to the whole set of activities that are not usually measured by traditional means (normal surveys) for economic or administrative reasons and NOE contains three components: Illegal activities, underground activities, and informal dealings. In rural areas, the number of persons employed in the informal sector was 96% compared to 74% in urban areas.

The 2008 distribution of employed persons in the sub-sector showed that Eastern and Northern Provinces had more persons engaged in forestry, fishing, and agriculture (Table 2.1). In the Copperbelt Province, employment in the formal mining sector accounted for 81.5% of persons while only 7.5% of people were employed in agriculture, forestry, and fishing. According to CSO (2004a,b), the informal employment accounted for more than 81% of the total population employed in

Province	Land Area (ha)	Area of Natural Forests (ha)	Area of Forests Plantations (ha)	Total Forest Area (ha)	% Employed in Agriculture, Forestry and Fishing
Central	9,439,438	7,910,249	236	7,910,485	12.8
Copperbelt	3,132,839	1,609,279	52,000	1,661,279	7.5
Eastern	6,910,582	5,152,294	2,843	5,155,137	19.1
Luapula	5,056,908	3,465,101	888	3,465,989	10.6
Lusaka	2,189,568	1,651,213	–	1,651,213	4.9
Northern	14,782,565	7,212,327	474	7,212,801	16.6
N-western	12,582,637	10,040,984	–	10,040,984	7.3
Southern	8,528,283	4,672,399	420	4,672,819	11.6
Western	12,638,580	8,254,154	–	8,254,154	9.5
Zambia	**75,261,400**	**49,968,000**	**56,861**	**50,024,861**	**100**

Table 2.1 Forest Area and Distribution of Employment by Province in 2008

Note: Figures of % employed in Agriculture, Forestry and Fishing include employees aged 15 years and above.

Zambia. At provincial level, the informal employment in Luapula Province—dominated in order by fishing> forestry> agriculture—had the highest number of people employed (95%) followed by Western Province with 92% of people employed and dominated in order by forestry> fishing> agriculture.

Since most of the rural communities are engaged in agricultural, forestry, and fishing activities for livelihood, statistics from these categories need to be broken down into lower units during surveys (CSO, 1998, 2004a). For example, the 1998 and 2004 surveys divided agriculture in strata of small, medium, and large agricultural activities then in lower units such as crops and livestock. This is important because the Government of Zambia fully recognizes the challenge to facilitate more broad-based wealth and job creation at even lower units. Policies have been developed towards private sector development as a strategic means to boost employment, with particular emphasis on the development of units such as Micro, Small, and Medium Enterprises (MSMEs) (GRZ, 2009). For example, the UN Green Jobs program introduced in 2013 aims to enhance the capacity and competitiveness of MSMEs in green construction, and in turn create 5,000 decent green jobs and improve the quality of 2,000 jobs in green construction (ILO, 2013). The program is being implemented by relevant ministries and institutions within the Government of Zambia and includes national stakeholders such as the National Association of Medium and Small-scale Contractors, Zambian

Association of Women in Construction, Zambia Institute of Architects, National Union of Building, Engineering and General Workers among others, with technical assistance from the ILO on business development services, UNCTAD on business linkages, UNEP on green building policy, ITC on financial services, and FAO on green building materials, products and technology, all with financial assistance from the Government of Finland.

According to the UN Green Jobs Program (ILO, 2013), there is a potential for job creation in the wood and wood products sub-sector particularly in the Zambian building construction industry. The program focuses on the value creation process linked to eco-friendly building materials and services within the building and construction industry, from production of raw material inputs such as sawn timber and wood-based products for green buildings to various end use applications. Due to its high labor-intensive nature, the wood and wood products industry offers excellent opportunities for broad-based job and wealth creation especially in (MSMEs).

According to the Forest Products Industry Opportunity Study conducted in Zambia in 2012 under the ILUA-FAO II project, the potential to double employment in this industry is very high considering the increasing demand for wood and wood products in the building industry. For example, carrying out reforestation and establishing forest plantations of about 2,500 ha of a *Pinus* species at farm level would require 10 foresters, 50 technical staff, 100 skilled staff and a number of general workers to be employed per year for 25 years. Similarly, the employment creation potential is enormous considering also the many sawmills currently operating in the country, employing over 10,000 persons (directly and indirectly) plus its employment multiplier effects in the downstream timber processing and service industry associated with value-addition manufacturing. Processing of kiln dried sawn timber into various value-added wood products would require a minimum of 25 workers per sawmill and an additional 5 workers to handle sales and marketing (Ratnasingam and Ng'andwe, 2012)

In 2012, the forest users associations in the softwood and hardwood-based sawmills recorded in excess of 1,052 sawmills operating in Zambia with the Zambia National Association of Sawmillers accounting for 340 sawmills. Copperbelt Sawmill and Tree Association had 400 sawmills while the Timber Producers Association and Lumber Millers Association of Zambia had 10 sawmills. This translates to 10,520 persons directly employed in timber processing.

PRODUCTION, MARKETS ACCESS AND TRADE OF WOOD PRODUCTS

The wood and wood products global trade is increasingly affecting local economies, governments and private organizations (Pepke, 2002; Banda et al., 2008). According to Pepke (2002) "Globalization" can be positive for consumers when it means more choice and lower prices for products, and vice versa when production is shifted to another region or closes down. Trade in wood, wood products, and paper products range from raw materials, such as logs, to finished products. Woodfuel (charcoal and fuelwood) have a growing importance in Zambia but local round-wood used for heating and cooking does not impact on global trade. However, government policies, around the world, in the effort to provide energy security and promote renewable energy sources in light of escalating fossil fuel prices have tended to create new international trade for woodfuels such as pellets and charcoal briquettes in value added forms.

OVERVIEW OF GLOBAL TRADE

The database for international trade is compiled by the United Nations Economic Commission for Europe (UNECE)/Food and Agricultural Organization TIMBER database, FAOStat, ForeStat database and UN Comtrade. The main importing and exporting regions of wood and wood products, globally, include North America, Latin America, Europe, the Commonwealth of Independent States (CIS), Asia, Oceania and China. According to Pepke (2002), the global export value of wood and paper products increased dramatically from 1996 to 2006, rising by over half (54.2%), from USD 132 billion to USD 204 billion (FAO, 2008). According to the statistics (FAO, 2008), global import values have risen less since 1996 (47.7%). There are many reasons why global imports and exports do not match, among which the most worrisome issue today is the trade of illegally harvested wood (Pepke, 2002) which is also quite problematic in Zambia.

The African region remains a minor player in international trade with most exports in the form of primary products such as round-wood, poles, and sawn wood (Pepke, 2002; FAO, 2008; Ratnasingam and Ng'andwe, 2012). Countries in the Oceania region have maturing plantations and thus have great potential in contributing to the global trade, but according to Pepke (2002) the growth of exports has been slow. For example, the rise in Chinese exports is quite recent, appearing within the last few years. On the other hand, Europe has seen the largest growth over the past decade, gaining $44.2 billion in exports and $102.2 billion in 2006 (Pepke, 2002; FAO, 2008).

According to FAO (2008) imports of wood, wood products, and paper products developed positively for some regions, especially Europe. Some regions have registered large percentage increases, but based on small values, such as in Africa, CIS, Latin America, and Oceania. The regional trends are important as they impact on domestic trade. For example, changing drivers of trade, both market and policy drivers, on a sub-regional basis, impacts on trade flows in many regions. According to Pepke (2002) considerable investment by western EU countries into central and eastern EU countries has been taking place to: (i) Gain access to wood resources; (ii) lower manufacturing and labor costs; (iii) access eastern European markets; (iv) take advantage of positive invest-ment climates; and (v) hire skilled employees. Relaxing border controls between EU countries has also facilitated the trade of wood and paper products in the EU region. Wood and wood products trade in Zambia is within the COMESA and SADC regions and faces similar challenges like that of Europe. In this grouping Zambia is the most strategically located sharing borders with many countries; being landlocked, it enjoys a comparative advantage to its neighbors.

WOOD AND WOOD PRODUCTS TRADE IN ZAMBIA

The manufacture of wood and products of wood is classified under Division 16 Section "C" of the ISIC. It includes the manufacture of wood products, such as sawn timber, plywood, veneers, wood containers, wood flooring, wood trusses, and prefabricated wood buildings. The production processes of these wood products include sawing, planning, shaping, laminating, and assembling. Trade in wood and wood products in Zambia in the past two decades (1960–1980) has been mainly associ-ated with copper production. According to Mostyn (1964), when copper production was 603,000 tons per annum, wood consumption was estimated at 283,000 m^3. The tree species used by the forest products industry during that period were Oregon pine sawn wood and eucalyptus poles (all imported) for mining and construction. Sawn wood from the Zambezi teak and *Brachystegia* species were used in the mines as pit props and development of the railways systems. During the period 2001–2010, copper production increased from 221,000 to 767,000 tons (CSO, 2005) with a concomitant increase in wood consumption estimated from 200,000 to 360,000 m^3 per annum, which is a rate of 0.47 m^3 of wood per ton of copper produced. Currently however, mining is no longer the direct driving force of wood consumption in the country, instead the domestic housing construction has gained that momentum and, to a lesser degree, the export demand from the neighboring countries has been steadily increasing (Ng'andwe, 2011b). Since the 1980s a number of wood products have been traded in various markets of Zambia with sawn wood traders dominating the domestic market (Figure 2.3).

In Zambia, nearly half of the domestic markets are involved in trading of wood products. Of the wood products traded in 2008, sawn timber had the largest share of actors (30.2%) in the market followed by wood furniture (16.6%) and charcoal (15.5%) (Figure 2.3). The trading process begins from logs that are cut into cants to sawn timber that may be processed further, or shaped by lathes or other shaping tools. The timber or other transformed wood shapes may also be subsequently planed or smoothed, and assembled into finished products, such as wood containers and wooden furniture. This group of items excludes kitchen cabinets and wardrobes which are made in combination with other non-wood materials.

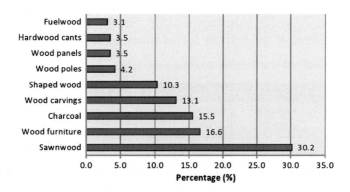

Figure 2.3 Percentage distribution of actors trading in different wood and wood products in domestic markets in Zambia.

PRODUCTION AND TRADE

Sawn Wood

The production of sawn timber in the country has been increasing from 114, 000 m^3 in 2001 to over 489,000 m^3 in 2010 (Table 2.2) and, generally, dominated by domestic consumption.

Imports and exports have been comparatively low and on yearly basis the apparent consumption of sawn timber (i.e., production + imports − exports) has been 315,000 m^3 (Table 2.2). The biggest market for sawn wood in the SADC region is South Africa even though the domestic markets demand over 90% of total production. In view of the timber shortage in the SADC region, exports of the manufactured products have been declining over years resulting in increased prices of round-wood and sawn timber within the region. There are indications that round-wood shortage will continue increasing in South Africa from 19.1% in 2010 to about 51% by 2030 (Crickmay et al., 2005). However, it is important that sawn wood and wood products statistics are reconciled from time to time to establish unaccounted movement of sawn timber. The local ENGOs could be used to compile and update such statistics in collaboration with the Central Statistics Office.

Wood-Based Panels

The level of WBP production and trade satisfies the domestic demand and is supplemented with modest imports of fiberboard which is not produced in Zambia. There are two companies producing particle-board, plywood, and blockboard that make up the WBPs. The FAO Forest Products Yearbooks (2012) reported a constant production of 18,000 m^3 per year and a decreasing trend on exports of WBP since 2001 (Table 2.3). The constant figures may be attributed to lack of national updates of the forestry sector statistics.

Table 2.2 Production and Trade of Sawn Wood in Zambia			
Year	Production (1,000 m^3)	Imports (1,000 m^3)	Exports (1,000 m^3)	Apparent Consumption (1,000 m^3)
2001	114	1	1	114
2005	353	1	6	348
2010	489	1	6	484
Average	318.6	1.0	4.3	315.3

Table 2.3 Production and Trade of Wood-Based Panels in Zambia				
Year	Production (1,000 m³)	Imports (1,000 m³)	Exports (1,000 m³)	Apparent Consumption (1,000 m³)
2001	18	7	0	25
2005	18	4	0	22
2010	18	3	3	18
Average	18	4.6	1	21.7

From FAO statistics, the falling consumption trend seems to be at odds with the undergoing construction boom in the country. Therefore, efforts must be made, particularly by the Forestry Department, to make available comprehensive annual reports which should include returns not only in monetary terms but also in volumes and quantities of forests, wood and wood products including WBP using standard units (Ng'andwe et al., 2008). In order to compute such statistics, conversion factors for wood and wood products are needed (Ng'andwe et al., 2008). In addition, considerable support from the government statistics is needed for the industry to grow and manufacturers to make necessary innovation that meets international standards.

The forest products industry in Zambia has a number of opportunities for enhancing marketing of wood products. To achieve this the Government vision 2030 and the sixth national development plans are aimed at establishing a profitable environment for increased domestic industrial growth, export promotion, the development of market-oriented production management and private sector development, business clusters and formation of economic zones. The Investment Act of 1993 offers a wide range of incentives in the forestry sector including special development allowances for growing certain crops, e.g., tea, coffee, banana plants, citrus fruit trees or other similar plants or trees.

Wood and wood products can be exported under the non-traditional products attracting associated incentives which provide for reduced income tax. In its policies, the government has made available to operators reduced corporate tax to 15% and special exemption from duty and sales tax on imports of machinery. Under manufacturing, the government has also listed a number of other opportunities and incentives available for further diversification, including areas of agro-processing, wood and leather products, food processing, textile production, and mineral processing in economic clusters. To assist manufacturing operations, the government established a Duty Drawback Scheme under which manufacturers producing export goods are entitled to claim a refund for duty paid back on raw materials used in the production process (SADC, 2006; GRZ, 1993, 2006b, 2011, 1993).

Fiberboard, Paper, and Paperboard Products

In Zambia, the fiberboard is imported from Zimbabwe and South Africa and only in small constant quantities (1,000 m^3) for many years since there is only one paper mill in Zambia—the Zambezi paper mills located in Ndola in the Copperbelt Province. The company produces about 4,000 MT of paper and paperboards from imported pulp (Table 2.4) and about 2,000 MT of printing and writing paper (Ng'andwe et al., 2012). Paper and paper products are classified in division 17 of section "C" in ISIC standard and includes the manufacture of pulp, paper, and converted paper products. Paper and paper products are grouped together because they constitute a series of vertically connected processes. To satisfy domestic needs, Zambia largely depends on imports for paper and paper products with the apparent consumption of over 31,000 MT per year.

Charcoal and Firewood

Production trends of charcoal and firewood since 2001 shows that woodfuel production increased from 7.9 million m^3 in 2001 to 9.1 million m^3 in 2010. According to the FAO Forest Products Yearbook (2010), Zambia produces an average of 1,041 MT of charcoal per

Table 2.4 (a) Production and Trade of Fiberboard and (b) Paper and Paperboard Products in Zambia

(a)	Fiberboard (1,000 m³)			
Year	Production	Imports	Exports	Apparent Consumption
2001	0	6	0	6
2006	0	1	0	1
2010	0	1	0	1
Average	0	2.6	0	2.6
(b)	Paper and Paperboard (1,000 MT)			
Year	Production	Imports	Exports	Apparent Consumption
2001	4	4	0	8
2006	4	27	0	31
2010	4	27	0	31
Average	4	19.3	0	23.3

Table 2.5 Production of Fuel Wood and Charcoal (2001–2010)			
Woodfuel Production (2001–2010)			
Year	Fuel Wood (1,000 m³)	Charcoal (1,000 MT)	Total Woodfuel (1,000 m³)
2001	2,469	1,041	7,945
2005	2,968	1,041	8,444
2010	3,643	1,423[a]	9,119
Average	3,026.6	1,168.0	8,502.6
[a]Estimate.			

annum (Table 2.5). These estimates have been constant over years and, apparently, they are underestimated as a large portion of the production from the informal sector is not recorded (Ng'andwe et al., 2012; Mwitwa and Makano, 2012).

It is estimated that a total of 5,476 m³ of round-wood (i.e., wood raw material equivalent) was used to produce 1,041 MT of charcoal based on the conversion factor of 5.26 m³ round-wood input per one metric ton of charcoal produced (Ng'andwe et al., 2012). The total woodfuel includes firewood and charcoal in cubic meters.

While fuel wood production increased over the years, there has not been a corresponding increase in charcoal production which is strange given the increase in local demand arising from reduced and erratic electricity supplies in both the urban and rural areas. Woodfuel is round-wood that is used as fuel for purposes such as cooking, heating or power production. Fuel wood is largely consumed in rural areas (68%), where the proportion of electricity usage as a source of cooking energy is only 2% (CSO, 2005, 2008b). On the other hand, charcoal is mostly consumed in urban areas where it represents about 51% compared to 42% electricity and 6% fuel wood (CSO, 2008a). The economic activity to produce woodfuel includes gathering and use of wood most of which is collected from indigenous forests. Woodfuel consists of the wood harvested from main stems, branches and other parts of trees and wood that is used for charcoal production (e.g., in pit kilns and portable ovens). The volume of round-wood used in charcoal production is estimated by using a factor of 5.6 to convert from the weight in metric tons of charcoal produced to the solid volume (m³) of round-wood input to production (i.e., 1.0 MT of charcoal = 5.26 m³ round-wood input) (Ng'andwe et al., 2012).

Charcoal production in Zambia increased from 649 MT in 1980 to 1,041 MT in 1993 and thereafter the production remained constant (Table 2.5), raising questions as to whether data on charcoal production are up-to-date (Mwitwa and Makano, 2012; Kewin, 2009; Ng'andwe and Ncube, 2012). However, recent findings show that charcoal consumption in Eastern, Copperbelt and Lusaka Provinces is over 1,423 MT (Mwitwa and Makano, 2012) while for the whole country it is estimated at 1,392 MT (Kewin, 2009). In modeling carbon dioxide emission reduction potential through the use of improved charcoal cook stoves in Zambia, Ng'andwe et al. (2012) found 1,153 MT of charcoal consumption instead of 2,511 MT that would be consumed if households continued to use traditional cook stoves.

VALUE-ADDITION OPPORTUNITIES

Tomorrow's wood products manufacturers will face distinctly different resources than are available today because global forests are being squeezed between growing needs and shrinking resources. Wood technology will play a catalytic role in finding solutions by increasing sawmill yields due to improved machinery; converting of by-products such as sawdust, slabs, bark, slabs, etc. to saleable products (e.g., particleboard; MDF and bark products) and recycling of paper and paper products. In recent years, new applications have appeared such as soil conditioning, compost, and landscaping. Today, huge volume of wood sawdust generated from processes is simply wasted. Therefore, developing by-products such as methanol, ethanol, and other chemicals will be a timely contribution to reducing the greenhouse gas (GHG) emissions.

In addition, wood products face competition on the market from other materials, such as plastics and metals, despite its superior eco-friendly advantages. One way of meeting these challenges is by value-added manufacturing at every stage of the supply chain including reducing costs of transportation and logistics. This section focuses on opportunities for Zambia and approaches to value addition that could work in the industry that is characterized by commodity type of wood products (Ratnasingam and Ng'andwe, 2012).

CHARACTERISTICS OF THE WOOD PRODUCTS INDUSTRY

There has been an industrial increase in round-wood production from 484,000 m^3 in 2001 to 1,813,000 m^3 in 2010, while woodfuel has increased from 7.94 million m^3 to 9.11 million m^3 during the same period (Ng'andwe et al., 2012). The characteristic of the wood products sector includes sawn wood which is predominantly untreated and used for construction and wood furniture (Ratnasingam and Ng'andwe, 2012). Exports have been minimal and often of inferior quality demanding remedial actions such as shown in Box 2.2.

To access international and regional markets, the wood products processing industries should be encouraged to embark on value addition and forest certification program. Zambia suffers from the "bountiful mentality" where the ample resources available negate the necessity to make full use of the resources.

APPROACHES TO VALUE ADDITION

The potential for value addition depends on several requirements being met in the Zambian wood products industry in a phased approach highlighted in Box 2.3.

This approach could be realized through taking several steps as follows:

1. Consistent quality of wood materials, in terms of moisture content, grades, legality status, treatment compliance must be met at all times.
2. Stringent regulatory framework that allows only "wood products of consistent quality" to be traded both in the domestic as well as the export markets must be enforced. This is the single most important aspect to boost customer confidence.
3. The institutions relevant to the wood products industry (Forestry Department, Bureau of Standards, etc.) must be strengthened. For instance, the Bureau of Standards of Zambia has 14 standards related to the use of wood in the construction, but its adoption in the market is limited. Aggressive awareness programs should be carried out to encourage the rapid adoption of such standards. Further, more standards should be formulated to cover the wood products in order to boost the confidence among wood products consumers.
4. Market sophistication must be increased, through collaboration between public–private partnerships. The lack of knowledge about wood materials especially among the consumers and specifiers (including the architectural fraternity) negates the need for any stringent

Box 2.2 Characteristics of the Wood Products Sector in Zambia

Domestic Market	Export Market	Remedial Actions
– Predominantly untreated sawn-timber, used for construction, BCJ, furniture, etc. – Transmission poles (treated with CCA and creosote) – Made-to-order wood products which are very price sensitive	– 1.5% of production, mainly as untreated or low quality sawn timber exported particularly to neighboring countries – 33% of panel products of international quality standards	– Enforcement and legislative framework for wood resource exploitation and processing is warranted – Strict quality codes and export procedures must be instituted to ensure value addition takes place – Grading rules, kiln-drying and treatment compliances must be met – Incentives to boost recovery of waste and mill residues needs to be formulated

Box 2.3 Phased Approach to Develop the Zambian Wood Products

Phase I: Quality Enhancement and Steady Volume	Phase II: Step Up Value Addition	Phase III: Business Skills
– Strengthen primary processing activities by increasing product quality, kiln-drying and treatment capacity and ensuring quality conformity – Enforce quality codes within the trade to boost product reliability – Timber extraction must be closely regulated to increase forest revenue, while ensuring future stocking – In order to enhance skills and capacity, cluster systems could be formed which will facilitate greater local community participation. For instance, the production of school furniture is one possible option for such initial phase efforts	– Strengthen secondary processing through the provision of incentives to use harvesting waste and mill residues (this will reduce pressure on plantation stocks)—this could be expanded to panel product producers – Public-Private-Financial Institution partnerships should be fostered to boost SME establishments, especially for waste collection and utilization – Boost research and development (R & D) activities in relation to the use of lesser known wood species and new products – Charcoal production based on waste material should be explored through the provision of incentives	– Process optimization to boost competitiveness through retraining of workforce and employment of skilled workers – Increase market sophistication through the provision of trade codes and standards that would guarantee quality and product conformity – Create awareness of green products through certification activities. For a start, public sector procurement of green wood products can provide the much needed price premium, which will pave the way for wider adoption of certification in the market

(Ramasingam and Ng'andwe, 2012).

regulatory measures to boost value added wood products manufacturing. As a result, the market is highly distorted towards low quality wood products.
5. Certification and legality of wood resources must be enhanced. Perhaps the participation of Zambia in the Forest Law Enforcement and Governance and Trade (FLEGT) initiative could be expedited to ensure that forest management gains greater governance.

The characteristics of wood products depict the pathways to greater value addition in the Zambian wood products sector. As illustrated, it is imperative that the R&D as well as product development practices are boosted in order to add value to the products produced. On the other hand, the market structure must be formalized to ensure a transparent distribution chain, in order to pave the way for greater value addition, and not to stifle product pricing. From the manufacturing practices, it is essential that good manufacturing practices are adopted within the Zambian wood products sector, in order to realize the value adding ambitions. Perhaps, kiln-drying, treatment, skilled workers training and quality control capacities must be the priority areas to realize market attractiveness and competitiveness of the sectors' products (Box 2.4).

Box 2.4 Matrix of Market Attractiveness and Competitive Strength

		COMPETITIVE STRENGTH	
	STRONG	**MODERATE**	**WEAK**
HIGH	**Extend Position** • Wood resources from natural forests	**Invest to Build** • Wood furniture • Builder CJ • Hardwood Components	**Build Cautiously** • Packaging • Paper • Paper boards
MEDIUM	**Build Selectively** • Wood-based • Panels • Pellets • Briquettes	**Invest Selectively** • Poles • Wooden Craft • General Utility Sawn wood	**Limit Expansion** • Charcoal • Fuelwood
LOW	**Protect and Refocus** • Forest Plantation Wood Resource • Private and Community Plantations	**Harvest** • Fuelwood	**Divest** ...

MARKET ATTRACTIVENESS

The following actions should be considered in order to realize market attractiveness and competitive strength shown in Box 2.1:

- Natural forest wood resources are most attractive for expansion, as the ample lesser-known wood species could be commercially exploited. Further, value could be further added through certification processes and application of FLEGT.
- Plantation wood resources must be strengthened especially in terms of reforestation, to ensure future sustainability of the dependent processing industries. Community forest plantation programs and private sector investment should be explored.
- Poles, craft, and general utility timbers must be protected through selective investment programs to minimize wastage. The general utility timbers usually find application in the construction sector, and through proper kiln-drying and preservative treatment, its value can be further enhanced and be used in wood furniture manufacturing.
- Investment into WBPs sector should be selective to correspond with availability of plantation wood resources, and also to utilize harvesting and mill waste. Briquette and pellets production could also be an option worth exploring to exploit the available waste.
- Furniture and builder's carpentry and joinery should be challenging due to lack of good manufacturing practices and lack of kiln dried timber. Investments should be selective on improving quality of products produced in the selected economic zones.
- The charcoal and fuelwood sector must come under closer control to ensure that the informality of the sector is brought under better management and regulation. This will allow the sector to be rationalized, at the same time release some of the valuable wood resource from the natural forest for industrial processing.

Through phased efforts, it would be possible to increase the value of the wood products currently traded in the domestic as well as in the export market. Estimates suggest that the domestic market for wood products in Zambia was valued at USD 225 million, while export was close to USD 12 million in 2010 (Ratnasingam and Ng'andwe, 2012). This is against an installed production capacity of almost 1.5 million m^3 per annum, suggesting that it would be possible to further expand the value adding potential through a phased approach as shown in Box 2.4.

In order to boost the potential of the wood products sub-sector, there is a need to institute a deliberate policy framework—legislative as well as fiscal instruments—to ensure that the wood resources are used efficiently and effectively. Box 2.4 provides a strategic map that could be adopted within the Zambian wood products sub-sector to ensure sustainable growth.

On the other hand, and apart from supplying wood fiber for processing, it would be possible to explore the benefits offered by the clean development mechanism (CDM) and REDD+, which could serve as a strong incentive for local communities to uptake proper forest resources management. Since afforestation and reforestation activities are a source of emission reduction, it is therefore imperative to gather baseline inventory data that would facilitate the adoption of the CDM and REDD+ schemes in Zambia to complete the value chain.

INTEGRATED USE OF FOREST AND MILL WOOD WASTE

The sawmill industry generates a lot of wood biomass and other residues that are left to waste. There are few companies that use sawdust and shavings as raw materials to feed the boilers for kilns and hot presses at WPIs (plywood, particleboard, sawmills, value-addition mills), Copperbelt Forestry Company (kilns), Bisonite (particleboard and kilns). On the other hand, small-scale producers simply burn the sawdust generated. However, some innovative sawmillers are using sawdust to produce compost manure, while others are using it as feedstock for wood sawdust cook stoves. Production of charcoal briquettes or pellets from wood sawdust is another potential project for the sawmilling industry in Zambia that requires modest investment. Value-addition processing of kiln dried timber has the potential to create employment downstream in wood furniture, construction and the building industry (Ratnasingam, 2007). In addition, value addition to all by-products, such as forest biomass for power generation and gasification, and use of sawdust in improved cook stoves among others, is considered highly important in the climate change mitigation debate (Ezzati et al., 2002; IIED, 2010).

Energy production from forest and mill waste as a renewable resource has become a topical subject since it is a renewable energy source. In general, energy is produced by burning woodfuel for cooking and heating. In recent years burning wood for fuelling steam engines and steam turbines that generate electricity has become common under emerging climate change strategies. At the household level various appliances ranging from three stone fire, furnace, cook stoves, fireplace, camp fire, etc., are used while industrial applications for power generation have emerged.

TYPES OF FOREST WASTE

There are various types of forest waste generated from land use change, such as industrial development, agriculture, mining, and construction (housing, roads, infrastructure, etc.). Wood remains the largest biomass energy source today; examples include forest residues (such as dead trees, branches and wood sawdust, yard clippings, and wood chips). Perhaps the most common forest and mill waste is firewood gathered from forests for household use. In general, all forest and mill waste are of biological origin thus considered biodegradable. This is in relation to either using harvested wood directly as a fuel especially at household level or industries in smelters and boilers by collecting from wood waste streams. For example, some sawmills have integrated use of mill waste to run boilers for kiln-drying of timber. In addition, the largest source of energy from wood is pulping liquor or "black liquor" a waste product from processes of the pulp, paper, and paperboard industry.

ENERGY PRODUCTION FROM WOOD WASTE

Biomass energy makes up 77% of the world's renewable energy or 10% of the world's total energy mix (3% in OECD and 22% in non-OECD countries) (IIED, 2010). It is the main source of energy in Malawi accounting for 97% total primary energy supply (Kambewa and Chiwaula, 2010). Biomass energy provides 68% of Kenya's national energy requirements and it is expected to remain the main source of energy for the foreseeable future (Mugo and Gathui, 2010). In Zambia it is estimated that 74% of cooking energy is from biomass energy, 16% from electricity and 10% from petroleum oil (Ng'andwe and Ncube, 2011). The woodfuel use is associated with the increase in population and is considered to be one of the drivers of deforestation (Mwitwa and Makano, 2012; Vinya et al., 2012). Forest waste biomass (logs, branches, leaves, twigs, chips, sawdust), as an energy source, can either be used directly via combustion to produce heat, or indirectly after converting it to various forms of biofuel. Conversion of biomass to biofuel can be achieved by different methods which are broadly classified into: *thermal*, *chemical*, and *biochemical* methods.

To reduce industry wood waste calls for a revamp and formulation of a framework that would not only reduce waste during harvesting and processing operations, but also make full use of the waste produced. Further, the licensing system and round-wood contracts must be re-examined to ensure that harvesting fee is calculated on volume-basis, and not on per tree basis as currently practiced, which will inevitably boost the forest revenue collected. To reduce waste from harvesting operations in plantations, forest plantation owners should conduct harvesting operations and simply sales round-wood from the landing areas. Another point of interest for harvesting round-wood from natural forests would be to encourage charcoal producers to exploit the waste from harvesting and processing operations, which could draw benefits from the CDM and REDD+ mechanisms. However, the Forest Department has a role to play by ensuring the proper framework is already on the ground and the local communities are aware of the benefits to be derived from such "waste to wealth" activities.

This process uses heat as the dominant mechanism to convert biomass into another chemical form. The basic forms of thermal conversion include combustion, pyrolysis and gasification. These forms differ principally by the extent to which the chemical reactions involved are allowed to proceed and mainly controlled by the availability of oxygen and conversion temperature.

Energy created by burning or combustion of forest biomass is also known as *dendrothermal energy*. This form of energy production is also suitable for countries where the forest waste is abundant or trees for fuel wood grows rapidly as is the case in some tropical countries such as Zambia. In recent years, the industrial application of thermal conversion technology has been the combined heat and power (CHP) generation and biomass co-firing with coal. This process generates both heat and electricity. In Tanzania such a plant has been operating using plantation grown timber while in Zambia the CHP plan has been a subject of investigation by potential investors in wood and energy sector using forest and mill waste (Figure 2.4). The integrated management of forest and mill wastes has been used during cement production and generation of electricity.

The process of thermo chemical decomposition of biomass at elevated temperatures in the absence of oxygen is known as *pyrolysis*. The word is coined from the Greek—derived *pyro* "fire" and lysis

Figure 2.4 (a) Forest and industry waste for dendrothermal energy production and (b) biomass briquettes as an example of fuel from sawdust.

"separating." This is normally the first chemical reaction that occurs during burning of wood and many organic materials. In a wood fire, the visible flames are not due to combustion of the wood itself, but rather of the gases released during the process of pyrolysis, whereas smoldering (flameless burning) is the combustion of the solid residue (char or charcoal) left behind by pyrolysis.

Charcoal is obtained by heating wood with a limited supply of oxygen and through complete pyrolysis—carbonization occurs, leaving only carbon and inorganic ash. In Zambia, charcoal is still produced in traditional earth kilns, by burning a pile of wood that has been mostly covered with mud or sawdust. The heat generated by burning part of the wood and the volatile by-products pyrolyzes the rest of the pile. The limited supply of oxygen prevents the charcoal from burning. There is another by-product generated from conversion of forest and agriculture waste, called biochar. Biochar is a name for charcoal when it is used for particular purposes, especially as a soil amendment. Like all charcoal, biochar is created by pyrolysis of forest and agriculture waste or virgin biomass. During land clearance for agriculture, plantation forestry or any land conversion activities, forests are cut generating waste that may be converted to charcoal for energy or biochar for soil amendment.

With so much forest waste being generated in some parts of Zambia, biochar is under investigation as an approach to carbon sequestration to produce negative carbon dioxide emissions through integrated management of agriculture and forests waste. Biochar is believed to have great potential in helping mitigate climate change via carbon sequestration, increase soil fertility and agricultural productivity. This is an area requiring extensive research since slash and burn has often been considered a major contributor to GHG emission and deforestation in Zambia (http://www. biochar-international.org/biochar).

Gasification

Gasification is an old technology, which flourished quite well before and during the Second World War, then disappeared soon after the Second World War. The gasification technology has undergone many ups and downs over the years. However, there is renewed interest in this old technology because of increased fuel prices and environmental concerns. It is now quite a modern and sophisticated technology with a decentralized energy conversion system which operates economically even for small-scale industries. Gasification is basically a thermochemical process which converts biomass materials such as forest and agriculture waste into gaseous components that can be used in various applications. The technology uses incomplete combustion inside the gasifier which yields producer gas, containing carbon monoxide, hydrogen, methane and some other inert gases. The produce gas is cleaned and purified so that it can be used to run gasoline or diesel engines with little modifications. The producer gas can also be used for the production of biofuels or for power generation through the firing of a gas engine, gas turbine or steam turbine boiler systems.

The design of gasifier depends upon types of fuel used and whether gasifier is portable or stationary. Complete gasification system consists of gasification unit (gasifier), purification unit and energy converter burners or internal combustion engine. There are many different kinds of gasifiers designs depending on the feedstock available. Portable gasifiers are mostly used for running vehicles while the stationary types are combined with engines and are widely used in rural areas of developing countries. They can be used for many purposes including generation of electricity and running irrigation pumps. Technologies such as biomass gasification which allow utilization of biomass fuel are of great importance in Zambia and these have been piloted in the rural areas of Ndola and Kaputa Districts. Theoretically, almost all kinds of biomass with moisture content of 5−30% can be gasified, however, not every biomass fuel can lead to the successful gasification. Most of the development work is carried out with common fuels such as coal, charcoal and wood.

Appliances Using Wood

A wood-burning stove is a heating appliance capable of burning wood-fuel and wood-derived biomass fuel such as pellets (Ng'andwe and Ncube, 2011). In general, most appliances consist of a solid metal closed fire chamber, a fire brick base and adjustable air control or simply air vent (Figure 2.5).

Traditional cook stoves are cheap but quite inefficient. According to Ng'andwe and Ncube (2012), there has been innovation in recent years to improve such cooks stoves as three stone fire and braziers to more improved models (Figure 2.6).

In the developed countries, various heating and cook stove appliances are used. Multifuel stove designs are common in the United Kingdom, Finland, Ireland and other countries around the World. Such appliances burn solid fuels only, such as firewood, wood pellets, coal, etc. They are typically made of steel or cast iron. Some models are also boiler stoves, with an attached water tank to provide hot water, and they can also be connected to radiators to add heat to the house.

Figure 2.5 Tradition appliances in Zambia: (a) Three stone cook stove and (b) traditional brazier. Source: Ng'andwe and Ncube (2012).

 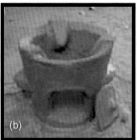 

Figure 2.6 Improved appliances: (a) Fixed mud stove, (b) portable clay stove, and (c) Pulumusa stove.

In a conventional stove, such as a traditional cook stove, when wood is added to a hot fire, a process called pyrolysis starts. Gases that evolved during this process are burned above the solid fuel. Air is, usually, admitted both below and above the fuel to enable complete combustion and to maximize efficiency. Many wood-burning stoves admit air above, due to the difficulty involved in getting the correct balance and in this case the volatiles are not completely burned and as a result energy is lost and consequently atmospheric pollution is unavoidable. Therefore, most improved cook stoves are designed to burn wood and regulate both fuel and air supply as opposed to controlling combustion of a mass of fuel by simple air flow as is the case in traditional stoves. For example, in a pellet stove, fuel is introduced into the pyrolyzing chamber with a screw conveyor and thus leads to better and more efficient combustion of the fuel.

FOREST CERTIFICATION AND GREEN BUILDING

There has been growing environmental awareness and consumer demand to promote environmentally appropriate, socially beneficial and economically viable management of the world's forest. Forest certification emerged as a credible tool in the 1990s for communicating the environmental and social performance of forest operations and sustainable forest management as a long term social, economic and environmental goals. A range of forestry institutions now practice various forms of sustainable forest management resulting in many potential users of certification, including: forest managers, scientists, policy makers, investors, environmental advocates, business consumers of wood and paper, and individuals.

It was during the 1990s that serious concerns about the state of World Forests and the inadequacy of intergovernmental processes and efforts to tackle deforestation gained momentum. One of the major concerns was deforestation which was linked to poverty issues especially in the tropical countries. Forest certification was an initiative which started following the United National Conference on Environment and Development in Rio de Janeiro, Brazil in 1992. Over the years, forest standards which include principles, criteria and indicators have been developed to evaluate the achievement of SFM at both the country and management unit level through an independent assessment of practices. This rise of certification led to the emergence of several different systems throughout the world including the FSC, The Program for Endorsement of Forest Certification

(PEFC), American Tree Growers Association, Sustainable Forest Initiative and others. As a result, there is no single accepted forest management standard worldwide, and each system takes a somewhat different approach in defining standards for sustainable forest management. Third-party forest assessments involve independent auditors who issue certificates to forest operations that comply with those standards developed by the respective scheme. Forest certification verifies that forests are well-managed—as defined by a particular standard—and chain-of-custody certification tracks wood and wood products and paper products from the certified forest through processing to the point of sale. In short, third-party forest certification is an important tool for companies seeking to ensure that the wood and wood products, paper, and paper-based products they purchase and use come from forests that are well-managed and legally harvested (Ng'andwe, 2003).

CERTIFIED FOREST PRODUCTS MARKETS

Markets for certified products have increased over the years in line with the increase for certified forest areas. For example, in 2006, certified forests supplied about 24% of the global industrial round-wood market (UNECE/FAO, 2008). FSC (2008) estimated annual sales of FSC labeled products at USD 20 billion. PEFC estimates that 45% of the world's round-wood production will come from certified forests by 2017 according to the State of the World Forest Report (FAO, 2009). In addition, to wood and wood products, other products are increasingly being certified, including woodfuel and NWFPs (UNECE/FAO, 2008). Both major certification systems now allow non certified wood to be sold together with certified wood under a "mixed sources" label, provided it meets certain basic requirements of acceptable forest management.

Since 1994 there has been more than 50 certification standards worldwide, addressing the diversity of forest types and tenures. However, globally, the two largest market based umbrella certification programs are the PEFC which is more or less a regional scheme and the FSC which is more or less international. The FSC scheme started in 1993 while the PEFC in 1999 and together with other schemes have a combined of area of over 350 million ha (Figure 2.7).

According to the Forest Products Annual Review for 2011–2012 period (UNECE/FAO, 2012) the FSC operated in 80 countries in 2012 and its certified forest area totaled 147.4 million ha compared to the

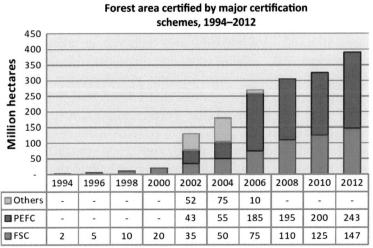

Forest area certified by major certification schemes, 1994–2012

	1994	1996	1998	2000	2002	2004	2006	2008	2010	2012
Others	-	-	-	-	52	75	10	-	-	-
PEFC	-	-	-	-	43	55	185	195	200	243
FSC	2	5	10	20	35	50	75	110	125	147

Note: *FSC means Forest Stewardship Council, PEFC means Program for the Endorsement of Forest Certification*

Figure 2.7 Certified forest areas and contribution from major schemes 1994–2012.

PEFC which had 243 million ha in the same period. In general, markets for certified forest products are in the environmentally aware customers in regions of Europe and North America. Therefore, according to UNECE/FAO (2012), certification, as market tool, cannot address all forestry concerns and it is likely that government regulations and other measures will continue to be necessary to address high risk situations. While recognizing the limits of voluntary certification and the role of government policy, making progress on the production of sustainable forest products will require a better integration of these roles if tropical deforestation is to be prevented. On the other hand, demand for wood products is still one of the main drivers of investment in sustainable forest management. Although short-term market changes influence individual decision-making, long-term changes in demand have a greater influence on investments in forestry and forest industry at the aggregate level.

The State of The World Forests Reports (FAO, 2014) brings out the following as key issues and trends in certification and markets for certified products:

• Although certification started with the objective of encouraging sustainable forest management in the tropics, only about 10% of the certified forest area is in the tropics.

- Certification provides access to markets where consumers prefer green products, but no price premium to cover the costs of certification. For many producers, access to the green market is insufficient incentive for seeking certification, especially when there is demand for comparable uncertified products produced at a lower cost.
- Major expansion in certification will depend on the response of consumers in rapidly growing markets (especially China and India).
- While the desire for market access may encourage the growth of certification, the main constraints could be on the supply side, especially the investments required to reach the minimum threshold level of management allowing certification.

GREEN BUILDING PRACTICES

Green building is new concept that is likely to add value to timber certification. "Green building" is a construction activity that conserves raw materials and energy and reduces environmental impacts. It refers to a structure and a process that is environmentally responsible and resource-efficient throughout a building's life-cycle: From siting to design, construction, operation, maintenance, renovation, and demolition. The Green Building practice expands and complements the classical building design concerns of economy, utility, durability, and comfort. Other related topics include sustainable design and green architecture.

Green building takes into consideration the future water use and energy demands, ecological site selection, and the procurement of sustainably produced materials. Some countries around the World including Zambia have started adopting green building standards and concepts. The Zambia Green Jobs Program is premised on a human and environment rights base as well as a value chain development approach for improving the sustainable livelihoods of rural and urban families through private sector development and sustainable housing (ILO/FAO, 2013). Like timber certification, it is also a third-party certification program that includes design, construction and operation of high-performance green buildings.

The aim of the Zambia Green Jobs Program is to enhance the competitiveness and sustainable business among MSMEs building construction sector. The Program seeks to unlock the job creation potential of the rapidly growing building construction sector in Zambia and has a competitive focus on the value chain for green building goods and services, from local production of environmentally friendly building materials through to more energy efficient building design. By so doing, the Program contributes to Zambia's Vision 2030, Sixth National Development Plan (SNDP), National Employment and Labor Market Policy, Decent Work Country Program, the MSME Policy as well as the sectoral private sector and financial sector reform programs funded by Finland and other Cooperating Partners. It forms part of the UN response through the United Nations Development and Assistance Framework in line with the aspirations of Zambia as a self-starter country for "Delivering as One" to increase the national ownership of UN activities.

The project targets a total of 3,000 MSMEs in the construction value chain from Central, Copperbelt, Eastern, Lusaka, Northwestern and Southern Provinces and includes architects, engineers, designers, wholesalers and retailers of green construction materials, medium and low cost housing contractors, sawmillers and timber growers. Working through their member associations and institutions, MSMEs become more competitive and effectively participate in the green building goods and services market and create more and better green decent jobs. The project was designed to work at three levels of the construction value chain, which are:

- *Micro-level*: MSMEs have enhanced capacity to penetrate the market for environmentally friendly building materials and services in Zambia.
- *Macro-level*: A refined policy, legal and regulatory framework for the building industry that stimulates uptake of green building materials and building designs.
- *Meta-level*: Increased stakeholder appreciation of business opportunities in green building construction.

CONCLUSION

Wood industry in Zambia has grown over the past several years and is now dominated by a number of income-generating outputs including sawn timber production, pole treatment, and WBPs. It provides over 10,000 jobs in both the informal and formal sectors. Various value-added wood products are produced and traded by large mills located mainly along the railway lines particularly in the Copperbelt and

Lusaka Provinces. There has been a modest export of sawn timber within the region—mainly to South Africa since 2001. Quality issues remain one of the major constraints for market access. Other major constraints facing the wood industry is lack of investment in new technology that maximizes the utilization and lack of value-added processing.

However, the opportunities exist to expand the wood industry through various pathways including:

- forest plantation establishment to supply short rotation wood fiber to industry under the Public Private Partnership;
- supply of lesser known hardwood species for value-added processing, builder carpentry and joinery, wood components and a wide range of end-use wood products;
- creation of kiln-drying facilities that will enhance value added processing in economic zones;
- integrated use of forest residues for value added products such as pellets, briquettes, fuelwood, wood charcoal and biochar;
- access to international markets through the application of the non VPA Forest Law Enforcement Governance and Trade, carbon trade and forest certification;
- creation of green jobs.

REFERENCES

Banda, M.K., Ng'andwe, P., Shakacite, O., Mwitwa, J. & Tembo, J.C., 2008. Markets for wood and non wood forest products in Zambia. Final Report Submitted to FAO-NFP, Liusaka, Zambia.

Campbell, B., 1996. The Miombo in Transition: Woodlands and Welfare in Africa. CIFOR, Bogota, Indonesia.

Chisanga, E.C., 2005. Establishment and status of forest plantations in Zambia. In: Ng'andwe, P. (Ed.), First National Symposium on the Timber Industry in Zambia. Mulungushi International Conference Center: Mission Press, Lusaka, Zambia, 29–30 September.

Crickmay, D.G., Brasseur, J.L., Stubbings, J.A., Daugherty, A.E., 2005. Supply and Demand Study of Softwood Sawlog and Sawn Timber in South Africa. Department of Water Affairs and Forestry, South Africa.

CSO, 1998. Central Statistical Office, Living Conditions Monitoring Survey Report. Zambia Printing Company, Lusaka, Zambia.

CSO, 2004a. Central Statistical Office, Living Conditions Monitoring Survey Report. Zambia Printing Company, Lusaka, Zambia.

CSO, 2004b. Central Statistical Office, Living Conditions Monitoring Survey Report. Zambia Printing Company, Lusaka, Zambia.

CSO, 2005. Central Statistical Office, Living Conditions Monitoring Survey Report. Zambia Printing Company, Lusaka, Zambia.

CSO, 2008a. Central Statistical Office, Gross Domestic Product 2005, Revised Estimates Real. Ministry of Finance and National Planning, Lusaka, Zambia.

CSO, 2008b. Central Statistical Office, Living Conditions Monitoring Survey Report. Zambia Printing Company, Lusaka, Zambia.

CSO, 2010. Central Statistical Office, Labor Force Survey 2008. Zambia Printing Company, Lusaka, Zambia.

Ezzati, M.B.M.M., Kammen, D.M., 2002. Comparison of emissions and residential exposure from traditional and improved cookstoves in Kenya. Environ. Sci. Technol. 34 (2), 578–583.

FAO, 2008. Food and Agriculture Organisation of the United Nations, FAOSTAT Database. Available at: < http://faostat.fao.org/site/626/default.aspx#ancor >.

FAO, 2009. Food and Agriculture Organization State of the World Forests Report, Rome, Italy.

FAO, 2010. Food and Agriculture Organisation of the United Nations Forest Products Yearbook 2010. FAO, ROME, Italy.

FAO, 2012. Food and Agriculture Organisation of the United Nations, Forest Products Yearbook 2012. FAO, Rome, Italy.

FAO, 2014. Food and Agriculture Organization State of the World Forests Report, Rome, Italy.

FSC, 2008. Forest Stewardship Council Principles and Criteria. Available at: < http://fsc.org/principles-and-criteria24.htm > (accessed 15.1.12.).

GRZ, 1965. Government of the Republic of Zambia, Forest Policy f 1965. Government Printers, Lusaka, Zambia.

GRZ, 1973. Government of the Republic of Zambia, Forest Act No. 39 of 1973. Government Printers, Lusaka, Zambia.

GRZ, 1993. Investment Act of the Government of the Republic of Zambia, Ministry of Finance and National Planning. Government Printers, Lusaka, Zambia.

GRZ, 1998. Government of the Republic of Zambia, Forest Policy of 1998. Government Printers, Lusaka, Zambia.

GRZ, 2006a. In: GRZ (Ed.), Government of the Republic of Zambia Vision 2030. Government Printers, Lusaka, Zambia.

GRZ, 2006b. Government of the Republic of Zambia, Zambia Development Agency Act of 2006. Government Printers, Lusaka, Zambia.

GRZ, 2009. Government of the Republic of Zambia, Public Private Partnership Act of 2009. Ministry of Finance and National Planning. Government Printers, Lusaka, Zambia.

GRZ, 2011. Government of the Republic of Zambia, Sixth National Planning and Development Plan_Final_Draft. Government Printers, Lusaka, Zambia.

IIED, 2010. Biomass energy—optimising its contribution to poverty reduction and ecosystem services. Report of an International Workshop, 19–21 October 2010, Parliament House Hotel, Edinburgh. IIED, London, UK.

ILO/FAO, 2013. Zambia Green Jobs Programme. Sustainable enterprises, creating more and better jobs. Lusaka, Zambia. Available at: <http://www.ilo.org/global/docs/WCMS_213390/lang–en/index.htm> (accessed 15.1.2015.).

ILO, 2013. Zambia Green Jobs Programme—building construction sector. Lusaka, Zambia. Available at: <http://www.zambiagreenjobs.org> (accessed 15.1.2015.).

Kambewa, P., Chiwaula, L., 2010. Biomass energy use in Malawi. A Background Paper prepared for the International Institute for Environment and Development (IIED) for an International ESPA Workshop on Biomass Energy, 19–21 October 2010. Parliament House Hotel, Edinburgh. Chancellor College, Zomba, Malawi.

Kewin, B.F.K., 2009. Carbon Stock Assessment and Modelling in Zambia: A UN REDD Programme Study. UNDP, FAO and UNEP, Lusaka, Zambia.

Masinja, A., 2005. Keynote address. Proceedings of the First National Symposium on the Timber Industry in Zambia 29–30 September. Mission Press Ndola, Zambia.

Mostyn, H.P., 1964. The Use of Forest Products by the Mining Industry of the Zambia Copper. Forest Products Research Bulletin No. 4. Forest Department, Ndola, Zambia.

MTENR, 2008. Integrated Land Use Assessment 2005–2008. Forestry Department, Ministry of Tourism, Environment, Natural Resources. Government Printers, Lusaka, Zambia.

Mugo, F., Gathui, T., 2010. Biomass energy use in Kenya. A background paper prepared for the International Institute for Environment and Development (IIED) for an International ESPA Workshop on Biomass Energy, 19–21 October 2010. Parliament House Hotel, Edinburgh. Practical Action, Nairobi, Kenya.

Mwitwa, J., Makano, A., 2012. Charcoal Demand, Production and Supply in the Eastern and Lusaka Provinces. Mission Press, Ndola, Lusaka, Zambia.

Ng'andwe, P., 2003. Timber Certification, Optimization and Value Added Wood Products—A Case Study for Zambia. Master of Science Forest Industries Technology, University of Wales, United Kingdom.

Ng'andwe, P., 2005a. Manufacturing value added wood products in Zambia. In: Ng'andwe, P. (Ed.), First National Symposium on Timber Industry in Zambia. Mulungushi International Conference Centre, 29–30 September, 2005. Lusaka, Zambia.

Ng'andwe, P. 2005b. Market access and the role of forest certification in Zambia. In: Ng'andwe, P. (Ed.), First National Timber Symposium. Mulungushi International Conference Center, 29–30 September 2005. Lusaka, Zambia: Mission Press, Ndola.

Ng'andwe, P., 2011a. Round Wood Supply and Demand Trends in Zambia and the SADC Region. Report Submitted to Zambia Forestry and Forest Industries Corporation. Copperbelt University, Kitwe, Zambia.

Ng'andwe, P., 2011b. Round Wood Supply and Demand Trends in Zambia and the SADC Region. Consultant Report Submitted to Zambia Forestry and Forest Industries Corporation, Ndola, Zambia. Copperbelt University, Kitwe, Zambia.

Ng'andwe, P., 2012. Forest Products Industries Development—A Review of Wood and Wood Products in Zambia. Submitted to the Ministry of Lands, Natural Resources and Environmental Protection and FAO, Lusaka, Zambia.

Ng'andwe, P., Ncube, E., 2011. Modelling carbon dioxide emission reduction through the use of improved cook stoves; a case for Pulumusa, portable clay and fixed mud stoves in Zambia, UNZA. J. Sci. Technol. 15 (2), 5–17.

Ng'andwe, P., Ncube, E., 2012. Modeling carbon dioxide emissions reduction through the use of improved cook stoves; a case for Pulumusa, portable clay and fixed mud stoves in Zambia. UNZAJST 15 (2), 5–15.

Ng'andwe, P., Muimba-Kankolongo, A., Shakacite, O., Mwitwa, J., 2006. Forest Revenue, Concessions Systems and the Contribution of the Forestry Sector to Zambia's National Economy and Poverty Reduction. FEVCO, Lusaka, Zambia.

Ng'andwe, P., Muimba-Kankolongo, A., J. Mwitwa, Kabibwa, N., Simbangala, L., Mulenga, F., 2008. Forestry sector guidelines for data collection and handling. Forestry Guidelines, Lusaka, Zambia.

Ng'andwe, P., Njovu, F., Kasubika, R., Chisanga, E., 2011. Forest Inventory of ZAFFICO Forest Plantations. Report Submitted to Zambia Forestry and Forest Industries Corporation. Copperbelt University, Kitwe, Zambia.

Ng'andwe, P., Simbangala, L., Kabibwa, N., Mutemwa, J., 2012. Zambia biennial compendium of Forestry Sector Statistics 1980–2010. Compendium, Ndola, Zambia.

Njovu, F., 2011. Socio-economic Study of ZAFFICO Forest Plantations. Report Submitted to Zambia Forestry and Forest Industries Corporation. Copperbelt University, Kitwe, Zambia.

Pepke, E.K., 2002. Global Outlook—Supply & Demand for Wood Products. Food and Agricultural Organization, UN Economic Commission for Europe, Geneva, Switzerland.

Puurstjärvi, E., Mickels-Kokwe, M., Chakanga, M., 2005. The Contribution of the Forest Sector to the National Economy and Poverty Reduction in Zambia. SAVCOR, Lusaka, Zambia.

Ratnasingam, J., 2007. The Malaysian Furniture Industry—A Pocket Guide. Asian Timber Publications, Kuala Lumpur, Malaysia.

Ratnasingam, J., 2012a. Estimated future potential for expanding value wood processing activities based on Zambian wood. Submitted to the Ministry of Lands, Natural Resources and Environmental Protection and FAO, Lusaka, Zambia.

Ratnasingam, J., 2012b. The Status of the Wood Products Sector in Southern Africa. IFRG Report No. 14, Singapore.

Ratnasingam, J., Ng'andwe, P., 2012. Forest Industries Opportunity Study—Synthesis Report Submitted to the Forestry Department Integrated Land Use Assessment II and the Food and Agriculture Organisation (FAO) of the United Nations Lusaka, Zambia.

SADC, 2006. Trade, industry and investment review. <http://www.sadcreview.com/country_profiles/zambia/zambia.htm> (accessed 20.10.2010.).

UN, 2006. The International Industry Classification of Economic Activities ISIC Revision 4 Statistical Papers. Agriculture, Forestry and Fishing. United Nations Department of Economic and Social Affairs Statistical Office, New York, NY, USA.

UNECE/FAO, 2008. Forest products annual market review, 2007—2008. Available at: <www.unece.org/timber> (accessed 20.10.2012.).

UNECE/FAO, 2012. Forest products annual review, 2011—2012. Available at: <http://www.unece/FAO.org/timber>.

Vinya, R., Syampungani, S., Kasumu, E.C., Monde, C., Kasubika, R., 2012. Preliminary study on the drivers of deforestation and potential for REDD+ in Zambia. A Consultancy Report Prepared for Forestry Department and FAO under the National UN-REDD+ Programme Ministry of Lands & Natural Resources, Lusaka, Zambia.

CHAPTER 3

Non-Wood Forest Products, Markets, and Trade

Ambayeba Muimba-Kankolongo[a], Phillimon Ng'andwe[b], Jacob Mwitwa[a], and Mathew K. Banda[c]

[a]Department of Plant and Environmental Sciences, School of Natural Resources, Copperbelt University, Kitwe, Zambia; [b]Department of Biomaterials Science and Technology, School of Natural Resources, Copperbelt University, Kitwe, Zambia; [c]Department of Computer Science, School of Mathematics and Natural Sciences, Copperbelt University, Kitwe, Zambia.

INTRODUCTION

Forests and woodlands harbor the country's rich biological diversity from which the rural communities directly derive their basic household's needs. Non-wood forest food products (NWFPs), such as fruits, vegetables, mushrooms and roots and tubers, which provide essential micronutrients mostly for children and women, are continuously harvested from the various indigenous forests for household's subsistence needs and for sale to earn income (Chidumayo, 1996; Campbell, 1996; Dewees et al., 2011). Edible insects—caterpillars and termites—and other secondary products such as honey and beeswax are all prized forest products that are currently in high demand in and outside the country. These resources, particularly food products and medicinal plants as well as several other products from grasses and fibres are widely traded in domestic markets in both rural and urban areas. The collection, processing, and trading of honey and beeswax, for instance, have reached an advanced stage of development in some locations of the country particularly in Northwestern Province where the activity now provides employment and substantial income to many households contributing positively to their livelihoods.

However, the existence of limited basic production statistics on NWFPs is of particular concern and deserves more attention considering their significant contribution to livelihood and poverty alleviation among the population (Puurstjärvi et al., 2005; Ng'andwe et al., 2006). The FAO supports this observation as the only available production data obtained from its statistical database are on honey and only up to

Forest Policy, Economics, and Markets in Zambia. DOI: http://dx.doi.org/10.1016/B978-0-12-804090-4.00003-3

1990 (FAO, 2003, 2005). Puurstjärvi et al. (2005) clearly demonstrated the contribution of the forest sector to the national economy and poverty reduction based on economic assumptions arising from previous available data on traded products as well as on their quantities and prices. In the following year, Mickels-Kokwe (2006) provided further insights on the marketing systems of some NWFPs in the country through the assessment of honey trading patterns. Broader investigations were then recommended to cover a range of other forest products including NWFPs to largely tap on the role of the forest sector to Zambia's economy and poverty alleviation. The end use distribution of NWFPs include for income generation, household consumption, and medicinal use (Figure 3.1).

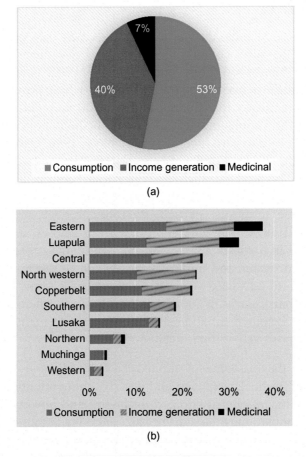

Figure 3.1 (a) End use distribution and (b) utilization of Non-Wood Forest Products in different Provinces of Zambia, 2006.

PREVALENCE AND USE OF NON-WOOD FOREST PRODUCTS

The harvesting and use of NWFPs in Zambia varies from Province to Province (Ng'andwe et al., 2006) and their importance within each Province also differs. For example, in the Southern part of the country, most hit by the recurrent drought, thatch grass and roots and tubers including yams are the most prevalent. They are collected by almost all households for both consumption as a relish and for sale as a source of income. Other important NWFPs in the area include the wild fruits, birds, mushrooms, leaves, and wild vegetables (Table 3.1).

Although the availability of the various forest products tends to be seasonal, their quantities throughout the country are currently dwindling as compared to the past (Mwitwa et al., 2012). The decline is mainly driven by unsustainable harvesting practices used by collectors, the increasing encroachment in most forest areas by settlers, the on-going devastating deforestation and in some cases by an increasing consumption trend particularly in the mining areas as well as in urban towns. For caterpillars for instance, countrywide spraying programme for tsetse fly control during the 1970s has been considered as one of the primary contributing factors to the reduction in their quantities more particularly

Table 3.1 Most Unutilized Non-Wood Forest Products in the Different Provinces of Zambia, 2007

Non-Wood Forest Products in Zambia		
Province	**Most Utilized**	**Others Utilized**
Southern	Thatch grass, roots, tubers and bulbs (*munkoyo, busala, chikanda*)	Wild fruits and vegetables, birds, mushrooms and leaves
Western	Mushrooms, honey, thatch grass, fruits	Medicinal plants, caterpillars, leaves, rattan
Luapula	Mushrooms, fruits, thatch grass, rattan, medicinal plants	Caterpillars, wild vegetables, bulbs, bults (*chikanda*), bamboos
Northern	Mushrooms, fruits, thatch grass, rattan, medicinal plants	Caterpillars, wild vegetables, bulbs, bulbs (*chikanda*), bamboos
Northwestern	Honey, mushrooms, thatch grass, wild fruits and vegetables	Caterpillars, medicinal plants, leaves, roots and tubers
Central	Roots, tubers and bulbs, thatch grass, mushrooms, wild fruits	Wild vegetables, leaves, caterpillars, mushrooms, honey
Lusaka	Roots, tubers and bulbs, thatch grass, mushrooms, wild fruits	Wild vegetables, leaves, caterpillars, mushrooms, honey
Eastern	Medicinal plants, thatch grass	Leaves, bamboos, caterpillars, roots, tubers and bulbs, wild fruits and vegetables
Copperbelt	Mushrooms, caterpillars, wild fruits, roots, tubers and bulbs, thatch grass	Wild vegetables, bamboos

in Southern, Western, and Northwestern Provinces. For mushrooms, the recurrent drought and/or loss of the forest cover as a result of deforestation as well as the destructive harvesting methods used are the main causes of the decline in their harvestable quantity.

NWFPs are collected in almost all the Provinces throughout the year creating varied levels of livelihood support. However, wild food crops, particularly the tubers and bulbs such as *chikanda* from orchids (*Satyria siva*), the traditional sweet beverage *munkoyo* processed from roots of the tree (*Rhynchosia heterophylla*), and wild yam known as *busala* (*Dioscorea hirtifolia*), which are used as food crops, medicinal plants, and cash crops have been nearly depleted from their habitat and their current distribution and availability are largely unknown (Mwitwa and Mulongwe, 2013). Some of these NWFPs such as the wild tubers and bulbs are cooked, roasted, or eaten raw. According to Mwitwa and Mulongwe (2013), the wild tubers and bulbs have seen a rise in commercialization in the last 20 years to an extent where natural stocks, particularly in some aquatic environments like dambos and river banks, have been depleted. The depletion is due mainly to unsustainable harvesting practices often involving uprooting of the entire plants. Continuous unsustainable harvesting practices have also affected the integrity of the natural environment resulting in the degradation of wetlands and in some cases contributing to increased greenhouse gas (GHG) emissions. Currently, the tuberous bulbs of *chikanda* (pseudo-bulbs), *munkoyo* and those in the *Dioscorea* genus constitute the major tubers that are widely consumed and traded in Zambia (Banda et al., 2008). Some of these tubers derive through informal cross-border trading between Zambia and DR Congo in Mwinilunga District in Northwestern Province.

In addition, several other tubers from species like *D. bulbifera, D. rotundata, D. cayenensis, D. dumetorum,* and *D. alata* are among the palatable food crops in the country (Muimba-Kankolongo et al., 2006), but others including those of *D. quartiniana* are mainly used for medicinal purpose due to the diosgenin hormone they contain which has aphrodisiac and energy boosting properties (Woodward, 2000; Mwitwa and Mulongwe, 2013). Generally, the tubers of the *Dioscorea* species are widely distributed throughout the sub-Saharan Africa (Coursey, 1967). Mwitwa and Mulongwe (2013) citing records from the East African Herbarium in 1952 reported that *D. bulbifera* and *D. alata* are widely cultivated as food crops and recommended that Zambia can promote the local community for domestication of more wild tubers and bulbs such as orchids and enhance their contribution to livelihoods which could reduce wetland and forest degradation. Furthermore, Mwitwa and Mulongwe (2013) emphasized for the need to conserve and domesticate more wild plants apart from a few that have been domesticated. They pointed out that the domestication has been largely unsuccessful due to lack of adaptable cultivation technologies coupled with lack of skills for the transfer of such technologies to growers. NWFPs contribute greatly to per capita incomes and the population health which includes improved household's nutrition particularly including addressing issues of HIV/AIDS (Barany et al., 2004; Mwitwa et al., 2013).

Limited information on production techniques for wild roots and tubers is foreseen as one of the major limiting factors for sustainable NWFPs domestication in the country. Muimba-Kankolongo et al. (2006) made collection of *busala* tubers from forests in Southern Province and attempted their cultivation and growth out of natural forests. However, Serenje (2002) tried to propagate *busala* using tissue culture techniques and found that the potential for propagation was immense although there was need to optimize the propagation media to induce shoot and root proliferation. Moreover, the National Irrigation Research Station in Magoye in Southern Province had also reported that the crop could be cultivated successfully from the rhizomes or roots (Mingochi and Sina, 2000). Profuse plant establishment was obtained by planting at the onset of the rains in October through December. The efforts should be revitalized, encouraged, and coordinated to ensure the sustainability of *busala* at local markets throughout the year. Such activities ensuring the growing and promotion of NWFPs on a large-scale, and the establishment of small agro-based processing and handling units for marketing of tuber products will enhance our understanding on their growth patterns and improve employment opportunities that could generate additional incomes at rural household levels. The benefits that might accrue from such development could include supplementing, diversifying, and enriching the nutrient value of the local food diet chains as well as the potential possibilities of increasing the income-earnings of the local people in the case of yams which have been domesticated in many parts of Zambia and the SADC region (Coursey, 1967; Falconer and Arnold, 1988; Mujaju, 2004; Phiri, 2003).

THE NON-WOOD FOREST PRODUCT INDUSTRY

NWFPs have been viewed as a means for a greater community self-sustaining for economic capacity development and are known to significantly contribute to household food security, incomes and general human health (Campbell, 1996; Chidumayo, 1996; Chidumayo and Siwela, 1988; Taulo and Mulombwa, 1998; Nair and Tieguhong, 2004). However, despite their potential for economic development, NWFPs have not been widely commercialized in Zambia. Moreover, national policies related to their production and use have been lacking. Although there are several reports on NWFPs in the country, their development, production and trade trends have in many cases been generally descriptive and area specific. During the 1990s, for example, (PFAP, 1998) established that most households in some areas in Central, Luapula and Copperbelt Provinces derived part of their livelihoods from various NWFPs. More importantly, a number of foods from forests such as mushrooms, *mufungo, munkoyo, chikanda,* honey, and caterpillars constitute an important source of household nutrition and also a safety net for the population during periods of drought and emergency foods in addition to providing incomes for other needs. Overall, PFAP (1998) provided well-documented and important information on the significant role NWFPs play in the livelihoods of poor households particularly in rural areas since they have less access to other sources of subsistence, income, and employment.

Nair and Tieguhong (2004) distinguished two models of commercial production of NWFPs in Africa in response to market demand. They include products which are collected from the forests, largely catering to either domestic markets or external markets; and the domestication of forest products to meet demands from domestic or external markets. According to Mtonga and Chidumayo (1996), there is greater potential for commercialization of NWFPs in Zambia. The increasing prices of most modern products including foodstuffs and medicines and the proximity of several rural areas to urban centres have offered great market opportunities for NWFPs (Mtonga and Chidumayo, 1996; Mulombwa, 1998; Nair and Tieguhong, 2004). Moreover, as indicated by FAO (2003), the high incidence of poverty and limited access to products and services of necessity from the formal sector has considerably contributed to the reliance on NWFPs for a large number of people, especially the poor in rural areas. Commercialization potential of NWFPs could also provide the collectors and sellers, especially women and children, with cash or through barter, additional goods. In Luapula and Northwestern Provinces for instance, Puurstjärvi et al. (2005) estimated the per capita cash income generated from sales of NWFPs by a total of 776 groups of people to range between USD55.41 and USD61.86. However, Mulombwa (1998) observed that the trading levels of NWFPs varied from community to community depending on the ethnic background, the product and its availability, distance from the market and accessibility to a reliable means of transport.

In many villages and small towns in the country, the contribution of forests and trees to food supply is essential for food security as they provide a number of important dietary elements that the normal agricultural produce does not adequately cover (Moore and Vaughan, 1994; Nkomesha, 1997; Mogaka et al., 2001; Guveya, 2006). Moore and Vaughan (1994) noted that gathered foods from forests such as caterpillars and mushrooms and their sales or exchange with other goods ranked higher than pension money or part-time employment and vegetables marketing in some parts of the Northern Province. Mogaka et al. (2001) also conclusively established that in several parts of Zambia, forest resources contribute significantly to the total income of poorer households but only a third of the income of the richer households. These observations were further substantiated by Puurstjärvi et al. (2005) and Guveya (2006). Using previously published data, Puurstjärvi et al. (2005) established through economic estimates gross added values for different forest products as their share to national GDP. Mushrooms, edible caterpillars, *chikanda*, and thatching grass constituted the main forest products from which the majority of the population obtained their income. Similarly, Nkomesha (1997) reported that selling of mushrooms, fruits, and *munkoyo* roots generated needed income to several households. But, Botolo (2003) had stressed the need for further research to link the economic growth and forest development in sub-Saharan Africa for the majority of poor people who derive their daily livelihood from forest-related economic activities (PRB, 2013).

Production, Markets, and Trade for Non-Wood Forest Products

The Miombo woodland in the country constitutes the major source of many food resources such as leaves, fruits and roots (Chidumayo, 1996), honey and mushrooms (Mwenechanya, 2003) and caterpillars, medicinal plants, and fodder (Taulo and Mulombwa, 1998; FAO, 2001). Several other important products from forests such as rattans, resins, gums, latex, tannins, colorants, ornamentals, and essential oils are also in high demand in the country (Njovu, 1993; Roper, 1997; MENR, 1997; Guveya, 2006). The importance of NWFPs in Zambia has been the subject of discussion locally and internationally (Mulombwa, 1998; Taulo and Mulombwa, 1998). Their utilization by most forest-dependent communities and their trading at both domestic and national markets represent an appropriate reason to enlighten about their potential and importance in people's lives in the country (Chileshe, 2001). Mwenechanya (2003) reported for instance that the Ababile Panini people of the Lufwanyama District in the Copperbelt Province utilized the highest variety of NWFPs, and fetched the highest income per household than other broad socioeconomic group-ings in the district. They derived income from the forest products such as fibre, mats, baskets, medicines, bush meat, wild foods, honey, and thatch-ing grass during almost all the months of the year. However, quantitative data on NWFPs production and utilization in Zambia is scanty as also highlighted from recent reports under the phase II of the Integrated Land Use Assessment project (Lwale and Gumbo, 2012).

MARKET ACCESS AND TRADE FOR NON-WOOD FOREST PRODUCTS

Although it is well documented that most NWFPs are used mainly for home consumption, some products such as basketry, honey and beeswax, caterpillars, mushrooms, and a variety of wild fruits are widely marketed in the country (Roper, 1997) and provide greater business opportunities for the rural communities (Chishimba, 1996; Ngoma, 2001). Trading of honey and mushrooms for instance has yielded millions of US dollars from both domestic and international markets for the country (Chishimba, 1996; Ngoma, 2001). Roper (1997) estimated that the combined value of traded NWFPs in Zambia exceeded the monetary value of timber mani-fold. In old Ming'omba village and in Chililabombwe on the Copperbelt, annual income generated from the sale of forest products was estimated to be USD836.00 per household out of which NWFPs accounted for USD724.00 (Ngoma, 2001). Based on economic calculations, Puurstjärvi et al. (2005) projected estimates showing the potential that various forest products have for their contribution to the national GDP.

Common Non-Wood Forest Products
The economic potential of some wild foods gathered from forests are undoubtedly vast. Collectively, wild foods add diversity and flavor to

various diets while providing also a food nutritional buffer rich in protein, energy, vitamins, and essential minerals to the most needy during seasonal and emergency shortages. As such, they constitute an important source of a considerable supplement to agricultural staple crops. They are an attractive income-generating activity for most rural people who are forest-dependent. Marketing of wild foods provides a huge opportunity for household incomes and the earned income is used to purchase other needed services. Poor households, in rural areas in particular, depend on forest foods for their livelihoods because they usually have more access to forests than to other resources (FAO, 1995).

Edible Vegetables and Fruits
Wild Vegetables
Several food diets in Zambia are mainly based on cereal and the starchy root and tuber staples which are supplemented, when possible, with other foods such as meat and fish that are high in protein. The forest leafy vegetables constitute a potential source for enrichment to such diets. Many of the edible forest leaves grow as weeds in cultivated and fallowed areas in forests (Vernon, 1983). He provided the most common forest herbaceous vegetables in the country including *Amaranthushybridus*, *Cleome gynandra*, *Corchorusolitorius*, *Sesamumaugustifolium*, *S. angolensis*, and *Bidenspilosa*. Others are *A. spinosa*, *A. thunbergii*, *Celosia trigyna*, *Portulacaoleracea*, *Cleome monophylla*, and *C. hirta*. However, Chidumayo (1996) indicated other numerous forest trees in the Miombo woodlands such as *Afzeliaquanzensis* and *Fagarachalybea* from which leaves are continuously harvested for use as vegetables. Wild vegetables such as *A. spinosa* and *B. pilosa* are available throughout the year while others are seasonally available. Most of these vegetables are usually collected in large quantities and are dried under the sun before or after being steamed and then stored. At times, they are dried, pounded and stored in powder form, thus making them convenient important sources of food during periods of food scarcity. Mulombwa (1998) reported that men, women, and children are all involved in harvesting of forest vegetables in Zambia.

Wild Fruits
Throughout the Miombo woodlands there are a very large number of tree species that bear edible fruits often consumed by Zambians. The use and trading of fruits are integral components of the local economies and culture, and play important roles in the household welfare in

Figure 3.2 Masuku fruits at Kasama Chikumanino Market in Northern Province, 2007.

every rural community. Often, important fruit trees are uncut in the fields during land preparation for agricultural crop production (Packam, 1993; Njovu, 1993). Chidumayo and Siwela (1988) listed about 82% of the 33 Zambian edible wild fruits with the greatest diversity being in the wet Miombo.

Among these, *Anisophyllea* spp., *Parinari curatellifolia, Sclerocarya birrea, Uapaca kirkiana* commonly known as masuku and *Zizyphus mucronata* trees produce the most commonly consumed wild fruits that are also even processed into beer or wine (Allan and Endean, 1966; Brigham et al., 1996). Moreover, they indicated that local utilization of the forest fruits in the country was often influenced by species distribution and abundance as well as flavor and tradition. These fruits contribute significantly to the diet of the population by providing the needed vitamins and minerals (FAO, 2001; Cori, 2003). Cori (2003) reported that in SADC countries, in general, the supply chain for marketing of wild fruits normally starts with farmers or local community members who collect fruits for home consumption and for trading (Figure 3.2).

Usually, fruits are sold at roadsides to other community members or to traders who in turn transport them to urban markets. Traders either function as wholesalers and sell to retailers, or directly sell fruits as retailers. Farmers or local community members at times sell their fruits directly to the public at markets, effectively eliminating all intermediary middlemen to negotiate for better prices. Because of their considerable importance, most of forest fruit trees are reserved under the Forest Law of Zambia (FAO, 2001; Muimba-Kankolongo et al., 2006).

Seeds and Nuts

Njovu (1993) recognized the importance of seeds and nuts as food in the country due to widespread distribution of a large number of tree species from which they are harvested. Trees such as *Adansonia digitata* and *Cajanus cajan* have seeds that are commonly traded in most of the drought prone locations of the country. However, because of scanty information available on the values and benefits that accrued from trading of seeds and nuts, Mulombwa (1998) advocated for more studies to establish the potential for their large-scale processing and marketing in the country.

Edible Bulbs, Roots, and Tubers

Orchids

The edible roots, tubers, and bulbs from the Miombo woodland are largely confined to herbaceous plants. FAO (2001) highlighted the importance of the edible roots, tubers, and bulbs from forests as crops for food security in Zambia especially in times of prolonged heavy rains or droughts. The most common and popular of the edible roots, tubers, and bulbs in the country are those of the orchids (*Satyria siva*)—commonly found in dambos within the Miombo ecosystem—from which a thick jelly paste *chikanda* is processed (Box 3.1), wild yam (*Dioscorea hirtifolia*) commonly known as *Lusala* or *busala*, and those of the legume *Rhynchosia heterophylla* out of which the sweet beverage *munkoyo* and *chibwantu* are made (Njovu, 1993; Chidumayo, 1996; Chishimba, 1996) (Figure 3.3). *Chikanda* is a well-known bulb product that can easily be prepared. The roots of *R. heterophylla* are also harvested and processed as an additive for use mostly in the production of *munkoyo* beverage drink.

Chishimba (1996) reported that trading of the products from these roots is lucrative. On the Copperbelt, about 150–200 g of *chikanda* could fetch as much as USD0.1. In Luapula Province, 300 g of the same product is sold at about USD0.22. IFAD through the MTENR, implemented a Forest Resource Management Project (FRMP) in which the orchids featured considerably in the whole of Northwestern Province (IFAD, 1999). One of the project objectives was to advocate for community participation in the management of forestry resources by promoting the sustainable harvesting, utilization, and propagation of NWFPs like orchids

Box 3.1 Recipe for the most popular bulb in Zambia

The most common and popular roots and tubers in Zambia are of the orchids from which a locally known African polony or sausage (*Chikanda*) is processed.

(a)

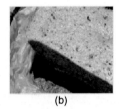
(b)

Recipe: Mash *chikanda* (a) bulb in a food processor or pound until fine. (1) Dry in the sun. (2) Bring water to the boil. Add sieved pounded nuts mixed with salt. (3) Mix in dried *chikanda* powder. (4) When mixture starts to thicken, add bicarbonate of soda and chilli powder mixed (if required). (5) Place the thick mixture in a refrigerator at room temperature until it sets (b). Serve sliced into wedges.

Figure 3.3 Processed roots of Rhynchosia heterophylla *being sold as an additive to* munkoyo *bevearage in Kasama, Northern Province, 2007.*

as alternative communities' income generating activity instead of forest destruction (Box 3.1).

Several communities in the Province embarked on the establishment of orchid gardens particularly in Salujinga and Kamafumbu areas where 650 and 800 rhizomes were maintained, respectively. As a result, the monetary value from harvesting and trading in orchid bulbs became more lucrative and important to the community's livelihood. The harvested orchids are sold at USD0.70 per kg to traders in urban area who resell them at Copperbelt markets for USD2.30 per kg.

Wild Yams

The wild yam, *Lusala* or *busala* in Tonga, occurs primarily on termite mounds (Chidumayo, 1996). It is considered as a food security and emergency crop with numerous comparative advantages in most of the drought prone areas of the country (Njovu, 1993; Chishimba, 1996). *Busala* is widely distributed across Zambia particularly in drought prone areas such as in Southern Province where it is either sold to earn income or eaten as a snack throughout the year (Muimba-Kankolongo et al., 2006) (Figure 3.4). They also reported other wild yam types, namely *Chipama* in Tonga and *Mupama* in Ngoni that are consumed only during periods of food shortage and the popular tuber crop commonly called Livingstone yam or Livingstone potato (*Plectranthus esculentus*) of the Family Labiatae (Woodward, 2000) in Western and Central Provinces.

Figure 3.4 Busala on sale by young girls in Northwestern Province, 2006.

According to Phiri (2003) and Mujaju (2004), it is a perennial herb producing edible roots commonly used in several Southern African countries. Some of these wild yams are widely cultivated, consumed in many households and sold either processed or unprocessed in markets and along the roads. Other important minor edible tubers which are also consumed in the country during drought periods include *Colocasiaedulis* and *Dolichoselipticus*(FAO, 2001). Obtained from forests by most rural people, *busala* is used for household consumption or it is sold at local marketplaces and to urban traders to sustain the livelihoods during periods of food shortages. It is an important traditional wild food for most households in drought prone areas of the southern part of the country (MAFF, 1999; Mingoch and Sina, 2000; MTENR, 2003a,b) where it is well sought after because of its importance in people's diets.

In some cases, individuals are just employed to find and dig the roots. It is used as a relish to accompany the local staple and is very popular among the Tonga speaking people of Zambia because of its reputation for treating menopausal related problems, as well as premenstrual syndromes in women (Ng'andwe et al., 2006). The methods of preparation depend on its availability and taste. It is first pounded and then groundnuts, tomato, or cooking oil may be added or also the tubers may be fried.

Other Non-Wood Forest Products
Mushrooms
During the rainy season, Miombo woodlands in the country produce up to 25 different species of edible mushrooms (Pegler and Piearce, 1980). Among the species of mushrooms most commonly used as food sources and for trading are *Termitomyces* sp., *Macrolepiota procera*, *Chanterelles* spp., *Schizophyllum commune*, *Russula* sp., *Lactarius* sp., and *Amanita zambiana*. These species contribute significantly to the overall daily lives of the rural people and considerably add value to the nutritional quality of their diets (Ingram, 2002). Mushrooms are an important dietary source of amino acids and income (Malaisse, 1997). They do not fruit at the same time and are therefore available at different periods during the rainy season (Figure 3.5).

Chanterelles (*Cantharellus cibarius*) consistently appear from December to February each year, but the availability of termite and Christmas mushrooms varies from year to year (Alien and Musenge, 1978; Piearce, 1978, 1981, 1982; Piearce and Francis, 1982; Högberg and Piearce, 1986).

Figure 3.5 Different species of mushrooms on sale at a roadside in the Copperbelt region, 2007.

Among these various species, only the *chanterelles* mushrooms are known to have international outlets for trading (Makano et al., 1996; Mulombwa, 1998). IFAD (1999) reported that Amanita Zambiana Ltd, one of the local private manufacturing companies, exported about 31.5 tons of mushrooms to Europe. Makano et al. (1996) estimated that about 2% of household income in Luansobe village in Masaiti District, Copperbelt Province, derived from the sale of mushrooms. According to Pegler and Piearce (1980), different kinds of mushrooms are displayed at one time or another of the year, representing species that are sufficiently well known and popular, and are gathered in large quantity for sale. Chishimba (1996) showed that trading of some forest products including mushrooms by local people on both national and international markets provides millions of dollars to Zambia. Generally, about one-third of rural households in the country harvest wild foods in the form of fruits, mushrooms, roots and tubers with a gross output of about 31 kg per household and that their roles in food security during famine periods were extremely important.

Honey and Bees Wax
Honey and bees wax are the earliest commercialized products among the NWFPs in Zambia (Mickels-Kokwe, 2006; Ng'andwe et al., 2006). Unfortunately, there is only limited precise statistics of honey and

beeswax production in the country. CSO, which is tasked with compiling national statistics, does not include beekeeping when undertaking agricultural or livelihood surveys. In 1992, about 90,000 kg of honey was reported to have been produced in Zambia with an estimated value of USD72,000.00 and during the same year, beeswax production reached about 29,000 kg, worth USD74,000.00 (MENR, 1997; Njovu, 1993). Honey-hunting from feral colonies of bees is done by many people on an opportunistic basis. Beekeeping, on the other hand, is often described as a "specialist enterprise," or a "way-of-life," originally practiced by only a small number of people with specific skills that are passed on from generation to generation within families.

Generally however, beekeeping utilizes the woodland ecosystem in two main ways including the partial domestication of wild bees by providing suitable places to establish bees colonies and through the hives made out of bark or wood that are provided to bees together with other equipment such as harvesting trays and ropes which are also made out of locally available resources. The traditional beekeepers often work within a framework set by the household's needs such as earning supplementary income, and the opportunity of supplying important commodities to the community for the enhancement of social relationships. Although, there is evidence suggesting that men are the most involved in this activity in Zambia, women from female-headed households are also increasingly taking the honey-hunting task to supplement their incomes (Clauss, 1991). FAO (2001) observed that beekeeping is best developed in Northwestern Province where about 6,000 beekeepers with about 500,000 hives produce over 600 t of honey and 100 t of wax each year (Figure 3.6). Outside Northwestern and Western Provinces, honey hunting is also prevalent in the rest of the country where hunters often destroy colony hives together with trees when fire is used to collect honey (MENR, 1997; Guveya, 2006). Guveya (2006) found that about 25% of households in some areas in Northwestern and Luapula Provinces are involved in honey production. Overall, they produce about 95 kg of liquid honey, 40 kg or comb honey and about 9 kgof wax which are for sale and often none of them having negative returns.

The need for cash can often result in the increase of the intensity of beekeeping, such as more frequent harvesting of honey and preparing value-added products due to market needs. Honey is of particular importance as an ingredient for honey beer (mead). In this form, it is

Figure 3.6 Honey from Mutanda in Northwestern Province sold in marketplaces throughout the country.

often regarded as a local currency amongst smallholder farmers in Northwestern Zambia where it is used for the payment of several other services such as hiring labor for field cultivation and for trading of cattle in some areas like Zambezi District (Clauss, 1991; Mulenga and Chizhuka, 2003). Similarly, Mulenga and Chizhuka (2003) estimated about 600–700 MT of honey being converted into honey beer annually.

In the Chiulukire local forest of Katete District in Eastern Province, the prices of honey often fluctuate considerably depending on its grading (Mutale, 2001; Mwape, 2002) (Figure 3.6). Mutale (2001) observed that producers in Mkokeza, Kankukute, Mutaya, Kolowela, and Mutopa villages distinguished the product into two categories, namely category one which is made up of Grade A of clean honey, Grade B of honey with pollen, and Grade C of honey with debris, that is, sundae ash, dead bees, etc.; and category two which is made up of Grade 1 (thick honey), Grade 2 (watery honey), and Grade 3 (honey mixed with pollen). Generally, clean and thick honey, which is ripe and sweeter, fetches more money compared to the pollen honey. In 2003, the total estimated production of honey in Zambia was at 1,500 metric tons of which 200 MT was traded within the country and some 250 MT exported to Europe (Ball, 2003). In 2004 however, Mickels-Kokwe (2006) estimated that the total value of the export product was more than 400 MT.

Because of the on-going deforestation arising from charcoal burning in Kaloko area of Masaiti District of the Copperbelt Province, beekeeping was emerging as a potential alternative source of income for rural communities than charcoal production (Kwenye, 2005). The annual income earned by a beekeeper (e.g., ZMW1 734 000.00) was considerably higher than that from charcoal production (e.g., ZMW663 250.00) and the time and energy spent for beekeeping were far less than expended in charcoal production.

Edible Insects

Generally, there are many edible insects of various types and values in different parts of the Miombo region (Lees, 1962; Malaisse, 1978; Chavunduka, 1975; Chidumayo, 1996; Vantomme et al., 2004). The most common edible insects in Zambia include the caterpillars and termite alates (Njovu, 1993) which are the winged termites. Young kings and the queens of Termite alates, especially of the *Macrotermes* spp., are captured and eaten as a source of protein and fat (Chidumayo, 1996; Chidumayo and Marjokorpi, 1997). Often, they leave their colony in the rainy season and during the nights, in a group or in a cluster of individuals, for a very short flight. They mate and then drop their wings looking for a new area to establish the nest, the so-called ant-hills. Mulombwa (1998) mentions grasshoppers being also one of the main sources of proteins for the rural population. These insects are mainly used as a food to provide for a cheaper and more accessible means of animal proteins especially for the rural households (Chavunduka, 1975; Mulombwa, 1998). The nutritional values of various caterpillar species are provided by Vantomme et al. (2004). Malaisse (1997) found that the average percentage of proteins and fats as well as energy of the 24 investigated fresh caterpillar species on a dry weight basis is 63.5($\pm$9.0%), 15.7($\pm$6.3%) and 457($\pm$32%) Kcal per 100 g, respectively. Species rich in calcium (e.g., *Tangoropsis flavinata*), in protein (e.g., *Imbrasia epimethea, I. dione, Antheua insignata*), or in iron (e.g., *Cinabra hyperbius*) are often prescribed for anaemic people particularly pregnant and breastfeeding women (Falconer and Arnold, 1988).

Insects from forests also constitute the main source of households' incomes during the periods when they are available (Figure 3.7). Chidumayo and Mbata (2002), in Northern Province, found that edible caterpillars generated incomes of over USD60.00 per household that are comparable to, or even higher than, incomes from the sale of

*Figure 3.7 Caterpillars (*Fikumbala*) and* chikanda *being sold at chilyapa market in Mansa District, Luapula Province, 2007.*

agricultural crops. The majority of the households in the area participate in caterpillar harvesting for both consumption and sale at a price
that would vary from year to year. For example, Chidumayo and
Mbata (2002) found prices around USD2.32 per kg of air-dried caterpillars. Although it was difficult to generalize on the income from
edible caterpillars in the Kopa area of Northern Province, the income
obtained from sales enabled the local people to buy other goods and
services that otherwise would be out of reach.

According to Mulombwa (1998), 1 kg of caterpillars, termites, and
grasshoppers is generally worth USD0.5. These insects are traded at
roadside markets closer to the points of collection often in rural areas
and at marketplaces in towns. Households that harvested enough insects
and caterpillars always generated more income, over the whole period
when the insects were available (Figure 3.8). Caterpillars are also bar-
tered or exchanged with other goods or services within the community
and grocery owners throughout the country where they are collected.

Bush Meat
Zambia's forests—game park reserves and natural forests—also constitute a rich reservoir of wildlife animal biological diversity. The country
has 19 national parks and 34 game management areas covering about
8.4% and 22% of the country's area, respectively. Many households

Figure 3.8 Grasses sold in Lusaka markets mostly for roofing of guesthouse shelters, 2007.

leaving around the Miombo woodland of varying ages consider these locations as important sources of various forest products including bush meat. This meat, which derives from a variety of wildlife animals, reptiles and birds, and bird eggs, constitutes an important source of protein for a substantial number of the population (FAO, 2001). Marks (1976) provides a detailed report on the existence of long history of traditional hunting for wildlife meat by the local people of Zambia and outlines how wildlife species became increasingly scarce during the late 1970s following an upsurge in commercial poaching (Marks, 1976). Similarly, Chidumayo (1996) had observed that the population of elephants had considerably diminished due to uncontrolled poaching. Traditional hunters are mostly men who hunt for both subsistence and cash needs. Several factors have contributed to the increasing reliance on natural resources like bushmeat for subsistence needs depleting the wildlife population. The most importants are the current increased population growth, the high level of poverty, and the widespread unemployment among the population (FAO, 2003). However, other small game like duiker and mammals (e.g., rodents and hares), and birds such as francolins and guineafowls are still widespread in the wild in the Miombo woodlands and constitute a common source of bush meat for subsistence needs in many parts of the country. Mulombwa (1998) reported that through the various forms of hunting licenses, the Government raises money, some of which is invested back into the management of national parks. The license fees for various animals often range from USD0.50 to USD391.00 for a Baboon and Sable/Roan antelopes, respectively.

Medicinal Plants

Numerous types of wild plants from forests including epiphytes, herbs, lianas and trees, and the different parts of plants such as roots, barks, leaves, flowers, fruits, and seeds are used in medicinal formulas and often prescribed by traditional healers as traditional medicines to cure several diseases. Traditional medicines are sold widely throughout the country (Cunningham, 1997) particularly in numerous herbal clinics existing all over the country where patients pay a lump sum of between USD1.00 to USD200.00. Moreover, Mulombwa (1998) observed that where modern medical services were limited or costly, most local people opted for forest flora or fauna as sources of medicines to treat diseases. However, there is much more to the trade in herbal extracts than records show because the trade is largely informal.

The general belief on traditional healing coupled with non-accessibility to and/or the deterioration of conventional medical services have led most people to search for traditional medicines to treat various illnesses (Chilufya and Tengräs, 1996; Cunningham, 1997). In Central, Luapula and Copperbelt Provinces, for example, forest-sourced medicines are mostly utilized by the majority of rural households (Nswana, 1996; PFAP, 1998). Similarly, Chilufya and Tengräs (1996) made reference to a wide knowledge about tree species used as medicines for treatment of various human ailments in northern Zambia. Forest medicines are regularly collected from about 80 different plant species mostly because of the poor distribution of rural health facilities, lack of money to acquire drugs and most importantly the cultural preference for traditional healing practices. Nswana (1996) as cited by FAO (2001) revealed that trading in traditional medicines in Central, Luapula, and Copperbelt Provinces was worth over USD43 million per annum. He observed that there is no gender and age bias in collection, processing and dispensing of herbal medicines though some cultural beliefs may impose temporary restrictions (e.g., for menstruating women or persons that are mourning).

In Mukuni village in Southern Province, mungongo tree (*Schinziophyton rautanenii*) is utilized by nearly 90% of the households for the selling of its products for their livelihoods (Mukela, 2005).

Table 3.2 Most Reputable and Commonly Used Tree Species as Medicinal Plants from the Miombo Woodland in Zambia

Tree Species	Plant Part	Medicinal Use
Aftezia quanzensis	Bark	Relieves toothache
Albizia antunesiana	Roots	Prophylactic against colds and coughs
Cassia abbreviate	Bark	Antibiotic
	Root extracts	Relieve toothache
Combretum molle	Leaf paste	Treatment of wounds and sores
Dichrostachys cinerea	Bark powder	Treatment of skin ailments
	Fresh leaves	Treatment of wounds and sores
Diospyros mespiliformis	Crushed roots	Ringworm treatment
	Crushed shoots	Treatment of wounds and sores
Diplorhynchus	Pounded bark	Wound dressing
condylocarpon	Forehead	Relieve headache
	Root extracts	Cough remedy
Garcinia huillensis	Aphrodisiac	Bark infusion
Hymenocardia acida	Vapor from boiling leaves	Relieves headache
Kigella Africana	Ripe fruits	Purgative
Lannea stuhlmannil	Leaf paste	Wound and sore dressing
Piliostigma thonningi	Chewed fresh leaves	Cough relief
Pterocarpus angolensis	Bark paste and/or ash	Treatment of skin ailments
Rothmannia whitfieldi	Unripe fruit juice	Treatment of wounds and rashes
Strychnos innocua	Seeds	Emetic properties

Mungongo was ranked as the major source of income in the village where as many males utilize the tree stems and roots for medicinal purpose as females, but more females use the fruits for consumption as males. Because of a very high demand for indigenous medicines, there have been concomitant increases in harvesting of the bark from some of the reputable tree species often leading to high prevalence of their death (Chidumayo, 1993; Geldenhuys and Mitchell, 2006). FAO (2000) estimates that Zambian annual exports of medicinal plants approximate about USD4.4 million. A summary list of common tree species that are utilized for medicinal purposes in Zambia as compiled by Banda et al. (2008) from Storrs (1982) is shown in Table 3.2. Of more than 100 tree species that have been documented, the frequency of use of the various plant parts ranges from 9% for fruits/seeds, 50% for leaves, 66% for bark, and 74% for roots.

Moreover, the bracket mushrooms (*Vanderbylia ungulata*) found on *Mubanga* trees (*Pericopsis angolensis*) is often sold by the local herbalists, claiming that it is effective in curing heart diseases (Sikombwa and Piearce, 1985). Addition of mushrooms into the diet might help in the hypocholesterolemic effect due to their dietary fibres such as β-glucans which increase intestinal motility and reduce the bile acid and cholesterol absorption (Ingram, 2002).

Fibre, Grass and Bamboos
As observed by Mulombwa (1998), information on the use of extractives and fibres in the country is barely available although it is common to find thatching in nearly all the rural areas. The business in grass trading has increased in recent years with the increasing construction of lodges for accommodation of tourists and other visitors (Figure 3.8). Besides providing household goods, fibres constitute an important source of dry-season income and an effective occupation for the rural communities.

Njovu (1993) reported that the most commonly used NWFP for basketry was the fresh canes of the bamboo (*Oxytenanthera abyssinica*). Bamboos have a wide range of uses both in rural and urban centres (Mulombwa, 1998; PFAP, 1998) including production of ropes for tying of bundles, fencing, making strings used mainly in hut construction, chair making, basketry, and mat making. Since 1989 to 1990s, there has been a flourishing of the handicraft industry using the indigenous bamboos in the Copperbelt Province (Chidumayo and Marjokorpi, 1997) and in Eastern Province (Mutale, 2001). Generally, the value creation from bamboos is often carried out near the selling points where baskets and other products are manufactured. Guveya (2006) indicated that bamboos production is active in Solwezi, Mansa, Kawambwa, Nchelenge, and Chiengi Districts where it is mostly performed by males. In Eastern Province, Mutale (2001) observed that various products such as winnowing baskets, chairs, tables, *mataza*, and shopping baskets are also made from bamboos, and their prices vary depending on the size and on whether they have been decorated or not.

Rattan and Other Non-Wood Forest Products
Rattan is also utilized to make mats, thatch and furniture. However, its only problem has been the limited resource base, as it does not often appear in many parts of the country (Mulombwa, 1998). Mutale (2001) found that trading of brooms from rattan was the most

Figure 3.9 Washing baskets, bamboo chairs, and mat reeds being sold at a specialized market between Mansa and Mwense Districts in Luapula Province, 2007.

lucrative activity in nearly all villages in Katete and Chipata Districts in Eastern Province. It was observed that the product is sold mainly by men while women constitute the bulk of buyers. The sellers can sometimes take them for selling as far as Lusaka (Figure 3.9).

Several other products such as resins, latex, tannins, essential oils, gums and dyes, and colorants are all important products deriving from forests. Njovu (1993) reported that edible essential oils are used widely though their production has been very low due to lack of appropriate processing facilities in the country. Although some of these products are used in combination with medicinal plants for various ailments, Mulombwa (1998) pointed out that they have not fully provided for their potential use on a commercial scale in the country mainly because of the lack of information on their specific use and market opportunities.

Production Trends and Marketing of Non-Wood Forest Products

Information regarding production and marketing patterns of NWFPs in the country is scanty (FAO, 2003; Ng'andwe, 2012). FAO (2003) reported that since NWFPs are produced and traded largely in the informal sector, there are few reliable statistics on their trends in production, consumption and trade. For instance, Table 3.3 shows production trends in honey production only from 1987 to 1992 as

Table 3.3 Trend in Honey and Beeswax Production from 1987 to 1992 in Zambia		
Year	Quantity (kg)	Quantity (kg)
1987	165,757	17,292
1988	180,782	14,765
1989	95,000	19,894
1990	205,305	56,395
1991	95,714	24,633
1992	90,000	28,000

compiled from Ministry of Environment and Natural Resources (1997) and Njovu (1993). Furthermore, Njovu (1993) observed that overall only about 25% of beeswax and 50% of honey production were recorded in official statistics. These official data estimated the total annual national production of honey during the 1990s as being more than 1,500 tons.

Generally, more than 50% of the honey produced in Zambia is exported to other countries, mostly Germany, Zimbabwe, South Africa, and Botswana (PFAP, 1998). Mickels-Kokwe (2006) outlined the overall honey marketing chain in the country. Although the structure of the chain is relatively simple, one may note however certain levels of vertical integration among Zambian honey buyers. He observed that the marketing pattern of honey is very simple and involves buyers who often try to take on processing, packaging, and distribution as well pricing to increase the profits. There are both domestic actors along the honey chain as well as the European players. For other NWFPs, Puurstjärvi et al. (2005) indicated that town-based traders procure the product from the rural supply areas, often camping in villages during collection periods. Local people collect products from the woodland and sell them fresh to traders for cash or in exchange with other goods. Then, the traders clean, process, and transport the products to town markets where they are sold at very profitable prices.

Production data compiled from Kambeu (2003), which are derived from sales of honey products to the Forestry Department Beekeeping Division, clearly demonstrate that although honey and beeswax production statistics have been consistent over years, the production patterns fluctuate considerably however from one year to another. Mickels-Kokwe

(2006) reported that fluctuations in the production are in part due to variations in the flowering of one of the main nectar species, the mutondo tree (*Julbernardia paniculata*). Several other factors such as variations in the rainfall pattern and lack of funds by the Forestry Department to purchase the whole product might have also influenced the fluctuation in the production trend. Furthermore, these data do not include the honey locally consumed or sold in local markets.

For products such as bush meat and medicinal plants, Nair and Tiegulong (2004) reported that the economic, social, and environmental underpinning their commercialization processes are completely different from subsistence. They are largely dominated by a network of intermediaries and most often they are market-driven resulting in their over-exploitation and depletion from forests. Bush meat and medicinal plant products, which are largely used locally, have become important products traded illegally to meet the domestic demand particularly of those who have migrated to urban cities. As a result of their increasing demand, there has been a concomitant decimation of a diversity of wildlife and tree species used as medicinal plants in certain areas in the country. For instance during the last few decades, bark harvesting for traditional healing has considerably increased due particularly to high levels of poverty among the population (Cunningham, 1997). In addition, the escalating high prevalence of HIV/AIDS and tuberculosis, and the greater number of traditional healers have significantly contributed to the increased demand for medicinal plants, putting considerable pressure on forests, to the extent that the survival of numerous tree species are threatened (Malambo et al., 2005). As a result of the proliferation of bark harvesters and traders, and thus increased competition, tree barks are harvested more regularly, often resulting in trees being completely ring-barked (Syampungani et al., 2005). Similarly, Malambo et al. (2005) observed that the high demand for bark has also lead to either the total stripping of some trees; complete felling of larger trees to get the bark from the entire length of the tree or the bark removal from juvenile trees.

MARKETS FOR NON-WOOD FOREST PRODUCTS

Although NWFPs are important in the daily livelihood of the majority of the population of Zambia, no concerted efforts have been devoted to develop a consistent yearly database for their production, utilization and trade. Apart from subsistence agriculture, the collection of NWFPs constitutes an important activity among the forest dwellers in the country mainly rural households, which they use for consumption and trade. However, as highlighted by Ng'andwe et al. (2006), the NWFPs subsector is fast growing and though its projected contribution to the Forestry sector GDP was lower, estimated at 0.04% in 2006, its economic importance links more to poverty alleviation rather than to the monetary gain. Furthermore, they established that the potential of various forest and woodland types to harboring a wide range of NWFPs varied considerably depending on their ecological characteristics and vegetation endowment. For some local community members, the forest based activity is the sole or principal source of income while for others, which is the highest group; this activity is undertaken occasionally mostly depending on the availability of some products with high cash-value and also as induced activity for the household's cash needs. This could be the need for school fees for the dependents, or for income to buy other household basic needs like foods and medicines, or to purchase other items such as agricultural inputs. Hence, the production of NWFPs per household and per capita varies considerably across households, villages, districts and Provinces. One of the motives for several household's increasing utilization of NWFPs is the alarming level of poverty among the population.

CSO (2008) clearly demonstrated that a large proportion of the population of Zambia lives in extreme poverty which is below the international poverty line of about USD1.00 per day. Malnutrition, due to poverty, afflicts considerably a significant portion of children, and infant mortality has been exceedingly high throughout the country (CSO, 1998, 2008). A glance at most children today and more specifically those in rural areas paints a rather unpleasant picture mainly because the majority of those under age 5 are malnourished (CSO, 2008). About 58% of children between 3 months and 5 years of age are classified as stunted, 25% underweight and 5% wasted. Infant mortality rate in the country is about 69 per 1,000 live births (PRB, 2013). Frequent illnesses and poor eyesight as well as increased maternal, infant, and child mortality constitute other forms of negative consequences resulting from under-nutrition in children in Zambia. To supplement family food and nutrient requirements, most households have reverted to forest foods as alternatives.

Most of the NWFPs are consumed directly at the household level after gathering, but some others constitute important mainstays of the family economy as they are mainly harvested for sale or barter in exchange with other products to satisfy basic households' needs (Box 3.2).

Box 3.2 Types of Non-Wood Forest Products regularly gathered in Zambia for household consumption and/or sale to earn income

Types	Products
Edible wild products	**Roots, tubers, and bulbs:** *Rhynchosia* spp., *Satyria siva, Dioscorea hirtifolia, Dolichos elipticus, Colocasia edulis, Plectranthus esculentus.*
	Vegetables: *Amaranthus* spp., *Cleome* spp., *Corchorus* spp., *Olitonius* spp., *Sesamum angolensis, Bidens pilosa, Celosia trigyna, Portulaca olenacea, Hibiscus* spp., *Adansonia digitat, Zanthoxylum chalybeum, Ziziphus abyssinica, Cratirociphon quarrei, Afzelia quanzensis, Agara chalybea.*
	Mushrooms: *Termitomyces* spp., *Macrolepiota procera,* Cantharellus spp., *Schizophyllum commune, Lentinus cladopus, Russula* spp., *Lactarius* spp., *Amanita zambiana, A. flammeola.*
	Fruits: *Anisophyllea* spp., *Parinari curatellifolia, Uapaca kirkiana, Sclerocarya birrea, Zizyphus Mucronata, Strychnos cocculoides, S. Spinosa, Vangueriopsis lancifolia, Garcinia huilensis, Syzidium* spp., *Mimusops zeyheri, Dalbergia nitida*
Bee Products	Liquid honey, comb honey, bee wax, propolis.
	Bamboo: *Oxytenanthera abyssinica*
Fibers; Construction materials	**Rattan:** *Calamus* spp.
	Thatch grass
	Wrapping leaves
	Reeds: *Phragnites mauritianus*
	Fibers: *Agave sisalan; Brachystesia* spp.
	Rafia: *Raphia fariniferi*
Medicinal products Extractive cosmetics and colorants Ornamental	Bark, leaves, roots, fruits/seeds, herbs
	Essential oils, dyes, gums, latex, resins, tannins, fats.
	Orchids, roots from trees used to make pots.

Trade in NWFPs is growing rapidly countrywide and it is common that almost every household, more especially in rural areas, is involved at one or the other period of the year in selling one or a variety of NWFPs. This activity is worth about USD56,000.00 of value-addition per annum in the country (Ng'andwe et al., 2006). Markets for NWFPs are growing mostly because of the extension of large urban centres, mainly along the line of rail, and the increasing scarcity of some forest resources that are now creating higher demands for several of these products. However, these income earning activities based on marketable forest products both in urban and rural areas could be seasonal or year-round depending on the products, or occasional when supplementary household cash income is most needed. Hence, the production of NWFPs for either domestic consumption or marketing is always done as part of the household's livelihood strategies such as securing foods and other essential subsistence goods for the family's social security (e.g., health care needs, education for dependents, household assets).

Traders at various marketplaces, in shops and along the road sides in Eastern Province order their products from surrounding villages and they use their own transport means each month to collect the products (Mutale, 2001). On selling, they increase the prices to cater for the cost of transport. Table 3.4 show the numerous types and local names of NWFPs that are produced, consumed and used in various ways.

Factors Constraining the Benefits of NWFPs

In Zambia, the sociocultural and economic benefits deriving from the NWFPs subsector are constrained by several factors including the biophysical, ecological, industrial, political, and economic factors.

Biophysical factors include over-harvesting of NWFPs mainly as a result of the increasing population growth together with lack of national efforts for reforestation that will ensure the expansion of the NWFPs-base in the country. As a result, the NWFPs total per capita income has been considerably affected due to the low quantities that are yearly traded.

Ecological factors include, for example, the impacts of climate change which has caused an increased mean of annual temperatures, localized flooding, changes in rainfall patterns, and recurrent droughts. These factors have resulted in the greatest negative impact on production of NWFPs such as mushrooms and fruits. Droughts and localized

Table 3.4 List of Local Names of Non-Wood Forest Products in Zambia

Categories	Products
Fibers and construction materials	• Bamboo (*Oxytenanthera abyssinica*) • Rattan (*Calamus* spp.) known as chiyenge • Thatch grass • Wrapping leaves • Bark from natural stands • Reeds (for basketry) • Fibers (*Agave sisalana*) used to make ropes • Rafia (*Raphia fariniferi*) or known as Ifibale

Vegetal	**Wild vegetables**
	  • *Amaranthus* spp. (Bondwe)
	  • *Cleome* spp. (Lubanga, Kabangaseeshe)
	  • *Corchorus* spp. (Lusaka)
	  • *Olitonius* spp.
	  • *Sesamum angolensis* (Mulembwe)
	  • *Bidens pilosa* (Kasokopyo)
	  • *Celosia trigyna* (Sunka)
	  • *Portulaca olenacea* (Cintekenteke Nyelele)
	  • *Hibiscus* spp. (Lumanda)
	  • Others (depending on local food preference)
	Leaves of tree species eaten as vegetable
	  • *Adansonia digitat* (Mubuyu)
	  • *Zanthoxylum chalybeum* (Pupwe chulu) *Ziziphus abyssinica* (Mukonga, Kangwa) *Cratirociphon quarrei* (Kafunda)
	  • *Afzelia quanzensis* (Mupapa, Musambamfwa)
	  • *Fagara chalybea* (Pupwe)
	  • Others (depending on local preference)

Wild fruits	**Trees**
	  • *Anisophyllea* spp. (Mufungo)
	  • *Parinari curatellifolia* (Mupundu, Mubula)
	  • *Uapaca kirkiana* (Mukusu or Masuku)
	  • *Sclerocarya birrea* (Muongo)
	  • *Zizyphus mucronata* (Kangwa, Kalanangwa)
	  • *Strychnos cocculoides* (Kasongole)
	  • *S. spinosa* (Sansa, Mupumangulube)
	  • *Vangueriopsis loncifolia* (Mungolomya)
	  • *Syzigium* spp.(Mufinsa)
	  • Others (depending on local food preference)
	Nuts
	  • *Parinari* spp. (Mufungo)
	  • *Schinziophyton rautanenii* (Schinz.) Radcl. -Sm. (Mungongo)
	  • *Ochna pulchra* (Munyelenyele)
	Seeds
	  • *Adansonia digitata* (Mubuyu)
	  • *Cajanus cajan* (Ingoliolio) or Pigeon peas
	  • *Guibourtia* spp. (Muzauli)
	  • Others (depending on local food preference)

Tubers, roots & bulbs	• *Rhynchosia* spp. (Munkoyo) • *Satyria siva* (Orchids) to make Chikanda • *Dioscorea hirtifolia* (Busala) • *Dolichos elipticus* (Chimbobolwa, Chilure) • *Colocasia edulis* • *Plectranthus esculentus* (Imyumbu) • Others (depending on local food preference)

(Continued)

Table 3.4 (Continued)	
Categories	Products
Mushrooms	Different species including: • *Termitomyces* sp., • *Macrolepiota procera*, • *Chanterelles* spp., • *Schizophyllum commune*, • *Russula* sp., *Lactarius* sp., • *Amanita zambiana* • and others
Bees' products[a]	• Honey • Bee wax
Edible insects	• Caterpillars (finkubala) • Termite alates (*Macrotermes* spp.) or Inswa • Cricket (Nyense) • Others (depending on local food preference)
Medicinal	• Bark • Leaves • Roots • Fruits/seeds • Herbs
Extractive cosmetics and colorants	• Essential oils • Dyes • Gums • Latex • Resins • Tannins • Fats
Ornamental	• Orchids • Tree roots used to make pots

flooding have the potential to destroy the population as well as the diversity of mushrooms resulting in either in yield decline or scarcity of the products during seasons when they are supposed to be available.

Industrial development in the country such as mining and its pull effects resulting in considerable urbanization also contribute to the reduction of the forest area available for the production of sufficient quantities of NWFPs. In some cases, forced relocation and restricted access rights for local communities to forest resources make NWFPs to be unavailable. In addition, air pollution and repeated contamination of the environment by leakages from mine tailings also have the potential to alter the soil pH, affect the reproductive patterns of pollinating agents for forest trees, and increase concentrations of heavy metals in soils and NWFPs, thus affecting their physiology and making them unsuitable for consumption. Moreover, clearing of large tracts of forests especially during site greening and development of urban centres removes a large quantiy of "tree and vegetation seeds" and changes the quality and extent of ecological habitats for growth of the next generation of NWFPs. Changes in the social patterns of local communities mainly due to the need for cash and urbanization may also influence the collectors' preference for either a salaried employment or engaging in activities that provide a constant and stable income.

The **government's development policies**, particularly those related to the promotion of nontraditional economy, have been largely biased against NWFPs (GRZ, 2011). Zambia relies on a copper- or mineral-dependent economy which has remained undiversified since 1964 (Fraser and Lungu, 2007). Surprisingly, NWFPs, such as *chikanda* that is now being sold as part of the cuisine of some hotels in the country, attract taxes for the same government as any other products. Therefore, reducing taxes on locally produced NWFPs could increase their consumption with a concomitant increase of the income accruing to rural communities.

The nonrecognition of the **economic value** of NWFPs, particularly their contribution to the national GDP and food security, constitutes a major drawback in terms of increasing their economic value in the country. National statistics reflecting their contribution to Zambia's GDP would reduce the "negative perception" policy makers have on NWFPs.

CONCLUSION

The forests and woodlands in the country constitute the major sources of different NWFPs including bushmeat, leaves, roots and tubers, fruits, honey, mushrooms, edible insects, medicinal plants, and fodders. Several other important products such as rattan, resins, gums, latex, tannins, colorants, ornamentals, and essential oils are all prized products that are in high demand. NWFPs play critical roles as households' safety-nets by providing food or income often in times of shortage and as important nutritional supplements especially for women and children. They are first hand opportunities for rural households to generate income and improve their living conditions. However, basic production statistics on NWFPs and information showing their contribution to poverty reduction and economy are scanty.

The informal segment of the national economy mainly uses NWFPs as one of the major sources of income, livelihood, and employment especially in the rural areas. Communities in these areas produce, consume, and market a great variety of NWFPs throughout the year. In areas with a high incidence of poverty, the consumption of NWFPs is more important than income generation from their sales and often, there areno surplus NWFPs available for sale while food shortages are common.

Generally, gathering of NWFPs in the country employs over 900,000 people accounting for 83% of the total human resource capital in the forestry sector. These are mostly either part-time employees or seasonal self-employed in the informal subsector. Honey harvesting alone has remarkably become a well-organized activity especially in the Northwestern Province where about 6,000 beekeepers with about 500,000 hives produce over 600 tons of honey and 100 tons of beeswax each year. Part of this production is exported, bringing some foreign exchange to the country. However, there is a general trend of a very little or no value addition involved in NWFPs trading, thus the products always fetch lower prices than should be the case. The role of NWFPs in supplementing the domestic economy and food security needs support from the public and private sectors through direct investment for sustainable forest resource management as a safety net for rural livelihood.

REFERENCES

Alien, A., Musenge, H., 1978. Studies on Cultivation and Utilization of Exotic and Indigenous Mushrooms in Zambia. Food Technology Research Ft2, Lusaka, Zambia, p. 9.

Allan, T.G., Endean, F., 1966. Manual of Plantation Techniques, ZAFFICO, Ndola. Forestry Department, Ministry of Lands and Natural Resources. Lusaka, Zambia (Mimeo). In: Endean, F. (Ed.). Ndola, Zambia.

Ball, D., 2003. Zambia. A Strategy for Developing production, Processing and Exporting in the Honey Sector. Report submitted to the Centre for the Development of Enterprise (CDE), Lusaka, Zambia.

Banda, M.K., Ng'andwe, P., Muimba-Kankolongo, A., Mwitwa, J., Tembo, J., 2008. Situation Analysis of Markets for Wood and Non-Wood Forest Products in Zambia. Analytical Report for the FAO (NFP) by BANKO Research Foundation and the Forestry Department — Ministry of Tourism. Environment and Natural Resources of Zambia, Kitwe, Zambia.

Barany, M., Hammett A.L., Stadler, K.E., Kengeni E., 2004. Non-timber forest products in the food security and nutrition of smallholders afflicted by HIV/AIDS in sub-Saharan Africa. Forests, Trees and Livelihoods, 14, 3–18.

Botolo, B., 2003. The value of forestry and forest products in the national income of the developing countries. In: Oksanen, T., Pajari, B., Tuomasjukka, T. (Eds.), Forests in Poverty Reduction Strategies: Capturing the Potential. EFI Proceedings No. 47. University of Helsinki, Finland, pp. 187–188.

Brigham, T., Chihongo, A., Chidumayo, E.N., 1996. Trade in woodland products from the Miombo region. In: Campbell, B. (Ed.), The Miombo in Transition: Woodlands and Welfare in Africa. Center for International Forestry Research (CIFOR), Bogor, Indonesia, pp. 136−174.

Campbell, B. (Ed.), 1996. The Miombo in Transition: Woodlands and Welfare in Africa, Vol. 5. CIFOR, Bogor, Indonesia.

Chavunduka, D.M., 1975. Insects as a source of protein to the African. Rhodesian Sci. News 9, 217−220.

Chidumayo, E.N., 1996. Handbook of Miombo Ecology and Management. Stockholm Environment Institute, Stockholm, Sweden.

Chidumayo, E.N., Marjokorpi, A., 1997. Biodiversity Management in the Provincial Forestry Action Programme Area. Forest Department, Ministry of Environment, Tourism and Natural Resources, Lusaka, Zambia.

Chidumayo, E.N., Mbata, K.J., 2002. Shifting cultivation, edible caterpillars and livelihoods in the Kopa area of Northern Zambia. Forests, Trees and Livelihoods, 12, 175−193.

Chidumayo, E.N., Siwela, A., 1988. Utilization, abundance and conservation of indigenous fruit trees in Zambia. Paper Presented at the ABN Workshop on Utilization and Exploitation of Indigenous and Often Neglected Plants and Fruits of Eastern and Southern Africa. Malawi, 21−27 August 1988.

Chileshe, A., 2001. Forestry Outlook Studies in Africa (FOSA) — Ministry of Natural Resources and Tourism — Zambia. Available Online: <http://www.Fao.org/Forestry/Fon/Fons/Outlook/Africa/Afrhom-E>.

Chilufya, H., Tengräs, B., 1996. Agroforestry Extension Manual for Northern Zambia. Regional Soil Conservation Unit (RSCU), Nairobi, Kenya.

Chishimba, W.K., 1996. In-Depth Study of Consumption and Trade in Selected Edible Vegetal Non-Wood Forest Products in Central, Copperbelt and Luapula Provinces. Provincial Forestry Action Programme. GRZ, Ministry of Environment and Natural Resources, Forestry Department; Department for International Development Co-operation, Ministry for Foreign Affairs of Finland, Ndola, Zambia.

Clauss, B., 1991. Bees and Beekeeping in the North Western Province of Zambia. Forest Department and Integrated Rural Development Programme (IRDP), Beekeeping Survey, Ndola, Zambia.

Cori, H., 2003. Background Report on Indigenous Fruit Commercialisation Activities in Selected SADC Countries. Commercial Products from the Wild Consortium, University of Stellenbosch, Matieland, South Africa.

Coursey, D.G., 1967. Yams. An Account of the Nature, Origins, Cultivation and Utilization of the Useful Members of the Dioscoreaceae. Longmans, Green and Co Ltd, London, United Kingdom, p. 230.

CSO, 1998. Central Statistical Office. Living Conditions Monitoring Surveys. Zambia Printing Company, Lusaka, Zambia.

CSO, 2008. Central Statistical Office. Living Conditions Monitoring Surveys. Zambia Printing Company, Lusaka, Zambia.

Cunningham, A.B., 1997. An Africa-wide overview of medicinal plant harvesting, conservation and health care. In: FAO (Ed.), Global Initiative for Traditional Systems of Health and FAO. Medicinal Plants for Forest Conservation and Health Care. FAO Non-Wood Forest Products Series No. 11. Rome, Italy, pp. 116−129.

Dewees, P., Campbell, B., Katerere, Y., Sitoe, A., Cunningham, A.B., Angelsen, A. et al., 2011. Managing the Miombo Woodland of Southern Africa — Policies, Incentives and Options for the Rural Poor, Program on Forests (PROFOR), Washington, D.C., USA.

Falconer, J., Arnold, J.E.M., 1988. Forests, Trees and Household Food Security. Social Forestry Network Paper 7a. Overseas Development Institute, London.

FAO, 1995. Valuing Forests: Context, Issues and Guidelines. Food and Agriculture Organization Forestry Paper No. 127. Department of Forestry, Rome, Italy.

FAO, 2000. Statistical Data on Non-Wood Forest Products in Africa: Draft Report. EC-FAO Partnership Programme. Forest Department, FAO, Rome, Italy.

FAO, 2001. Non-Wood Forest Products In Africa: A Regional and National Overview. Working Paper Fopw/01/1. Forestry Department, Food and Agriculture Organization, Rome, Italy.

FAO, 2003. Forest Outlook Study for Africa: Sub-regional Report—Southern Africa. Development Bank, European Commission and FAO, Rome, Italy.

FAO, 2005. Global Forest Resources Assessment. United Nations Food and Agriculture Organisation, Rome, Italy.

Fraser, A., Lungu, J., 2007. For whom the windfalls? Winners and losers in the privatization of Zambia's copper mines. Available at: < http://www.minewatchzambia.com >.

Geldenhuys, C., Mitchell, D., 2006. Sustainable harvesting technologies. In: Diederich, N. (Ed.), Commercialising Medicinal Plants—a Southern African Guide. Sun Press, Stellenbosch, South Africa, pp. 21–40.

GRZ, 2011. Zambia Six National Development Plan 2011–2015: Sustained Economic Growth and Poverty Reduction. Ministry of Finance and National Planning. Government Printers, Lusaka, Zambia.

Guveya, E., 2006. Final Draft Report of the Mid-Term Baseline Survey for Forest Resource Management Project (FRMP). Ministry of Tourism, Environment and Natural Resources (MTENR). Forestry Department, Lusaka, Zambia.

Högberg, P., Piearce, G.D., 1986. Mycorrhizas in zambian trees in relation to host taxonomy, vegetation type and successional patterns. J. Ecol. 74, 775–785.

IFAD, 1999. International, Funds for Agricultural Development (IFAD) Report on Forest Resource Management Project, Report No. 1232-Zm. Rome, Italy.

Ingram, S., 2002. The Real Nutritional Value of Fungi. Available at: <http://www.davidmoore.org.uk/Sec04_12.htm>.

Kambeu, K., 2003. An overview of beekeeping development in Zambia. In: Environmental Conservation Association of Zambia (ECAZ) (2003b). Proceedings of the national beekeeping association of Zambia workshop held at Barn Motel, Lusaka, 10 June 2003. Smallholder Enterprise and Marketing Programme (SHEMP), Lusaka, Zambia.

Kwenye, M.J., 2005. Beekeeping: An Alternative to Charcoal Production as a Source of Income for the Local Community in Kaloko Area. A Special Project in Partial Fulfillment for the Award of A B.Sc. Degree in Forestry. School of Natural Resources, Copperbelt University, Kitwe, Zambia.

Lees, H.M.N., 1962. Working Plan for Forests Supplying the Copperbelt and Western Provinces. Forest Department. Government Printers, Lusaka, Zambia.

Lwale, M., Gumbo, D., 2012. Measuring the Informal Forest-Based Economy as Part of the National Forest Monitoring and Assessments — Progress Report No. 2. Submitted to ILUA II Project. Lusaka, Zambia: MTENR.

MAFF, 1999. Proceedings of the Annual Planning Meeting for Crop Improvement and Agronomy Division, Department of Research and Specialist Services Soils and Crop Research Branch. Mount Makulu Research Station, Chilanga, October 1999.

Makano, A., Ngenda, G., Njovu, F., 1996. The Contribution of Forestry Sector to the National Economy. A Task-Force Report for ZFAP. Department of Forestry, Ministry of Environment and Natural Resources, Lusaka, Zambia.

Malaisse, F., 1978. The miombo ecosystem. In: UNESCO/UNEP/FAO (Ed.), Tropical Forest Ecosystems, a State of Knowledge Report. UNESCO, Paris, France, pp. 589–606.

Malaisse, F., 1997. Se Nourrir En Forêt Claire Africaine. Approche Écologique Et Nutritionnelle. Les Presses Agronomiques De Gembloux, Gembloux, Belgique.

Malambo, F.M., Mbunda, J., Mitchell, D., 2005. Indigenous knowledge and best bark harvesting practices: a local community perspective. Proceedings of the SADC Regional Workshop for "Trees For Health Forever": Implementing Sustainable Medicinal Bark Use in Southern Africa. Willow Park, Johannesburg. 1–3 November 2005, pp. 1–5.

Marks, S.A., 1976. Large Mammals and a Brave People: Subsistence Hunters in Zambia. University of Washington Press, Seatle, WA.

MENR, 1997. Zambia Forest Action Plan (ZFAP), Vol. II. Challenges and Opportunities for Development. Ministry of Environment and Natural Resources. Government Printers, Lusaka, Zambia.

Mickels-Kokwe, G., 2006. Small-Scale Woodland-Based Enterprises with Outstanding Economic Potential: The Case of Honey in Zambia. CIFOR, Bokota, Indonesia.

Mingochi, D.S., Sina, W.S.L., 2000. Improved Vegetable Production Practices for Smallholder Farmers in Zambia: A Reference Manual for Field Extension Workers. Ministry of Agriculture, Food and Fisheries (MAFF), Lusaka, Zambia.

Mogaka, H., Gasheke, S., Turpie, J., Emerton, L., Karanja, F., 2001. Economic Aspects of Community Involvement in Sustainable Forest Management in Eastern and Southern Africa. Forest and Social Perspectives in Conservation No. 8. IUCN-Eastern Africa Regional Office, Nairobi, Kenya.

Moore, L., Vaughan, M., 1994. Cutting Down Trees: Gender, Nutrition and Agriculture Change in the Northern Province of Zambia, 1890–1990. James Currey/Porthmouth Heinemann, NH. London, UK, p. 278.

MTENR, 2003a. Ministry of Tourism, Environment and Natural Resources, Joint Forest Management Plan for PFAP Il: Local Forest. Forest Department, Namwala District, Zambia.

MTENR, 2003b. Ministry of Tourism, Environment and Natural Resources, Joint Forest Management Plan for Ndondi Local Forest. Forest Department, Choma District, Zambia.

Mtonga, B., Chidumayo, E.N., 1996. The Role of Women in Forestry in Zambia. ZFAP Secretariat, Lusaka, Zambia.

Muimba-Kankolongo, A., Njovu, F., Boby, S., 2006. Baseline Survey of Cultivation, Utilization and Marketing of Yams in Zambia. A Survey Report to Copperbelt University (CBU), Jambo Drive, Riverside. Kitwe, Zambia.

Mujaju, C., 2004. An Ecogeographic Study of Selected Vegetatively Propagated Plants Occurring in Zimbabwe. SADC Plant Genetic Resources Centre, Lusaka, Zambia, p. 78.

Mukela, I., 2005. The Potential of Mungongo in Improving the Socio-Economic Status of Rural Communities: A Case Study of Mukuni Village. A Special Project in Partial Fulfillment for the Award of B.Sc. Degree in Forestry. School of Natural Resources, Copperbelt University, Kitwe, Zambia.

Mulenga, A.M., Chizhuka, F., 2003. Industry Profile of Honey in Zambia. Lusaka, Zambia (Mimeograph).

Mulombwa, J., 1998. Non-Wood Forest Products in Zambia. A Report of the EC-FAO Partnership Programme (1998–2000). Project Gcp/Int/679/Ec Data Collection and Analysis for Sustainable Management in ACP Countries — Linking. National and International Efforts. Forest Department, FAO, Rome, Italy.

Mutale, G., 2001. Contribution of Selected Non-Timber Forest Products to the Livelihood of Local Communities: A Case of Chiulukire Local Forest. Katete, Zambia. A Special Project for Partial Fulfillment of B.Sc. Degree in Forestry. School of Natural Resources, Copperbelt University, Kitwe, Zambia.

Mwape, W., 2002. Beekeeping in Zambia. A Publication Intended for Rural Beekeepers in Zambia. Forestry Department, Ministry of Tourism and Natural Resources, Lusaka, Zambia.

Mwenechanya, C., 2003. Forestry, Poverty Alleviation and Sustainable Development: A Case Study of the Lufwanyama District, Copperbelt Province. A Special Project in Partial Fulfillment for the Award of B.Sc. Degree in Forestry. School of Natural Resources. Copperbelt University, Kitwe, Zambia.

Mwitwa, J., Mulongwe, L., 2013. Community Based Forest and Wetland Resources Management: Domestication, Commercialisation and Utilization of Wild Tubers and Bulbs in Central Province. School of Natural Resources, Coppebelt University, Kitwe, Zambia.

Mwitwa, J., Muimba-Kankolongo, A., German, L., Puntodewo, A., 2012. Copper Mining, Forest Management and Forest-Based Livelihoods in the Copper Belt of the Democratic Republic of Congo and Zambia. LAP LAMBERT Academic Publishing GmbH & Co. KG. German.

Mwitwa, J., Mulongwe, L., Ng'andwe, P., 2013. Community Based Forest and Wetland Resources Management — Domestication, Commercialisation & Utilisation of Wild Tubers & Bulbs Project Document — School of Natural Resources. Copperbelt University, Kitwe, Zambia.

Nair, C.T.S., Tieguhong, J., 2004. Africa Forests and Forestry: An Overview. Report on Lessons Learnt on Sustainable Forest Management in Africa. FAO, Forestry Department, Rome, Italy.

Ng'andwe, P., 2012. Forest Products Industries Development — A Review of Wood and Wood Products in Zambia. Lusaka, Zambia: Submitted to the Ministry of Lands, Natural Resources and Environmental Protection and FAO, Lusaka, Zambia.

Ng'andwe, P., Muimba-Kankolongo, A., Shakacite, O., Mwitwa, J. 2006. Forest Revenue, Concessions Systems and the Contribution of the Forestry Sector to Zambia's National Economy and Poverty Reduction. FEVCO, Lusaka, Zambia.

Ngoma, J., 2001. The Potential of Forestry to Contribute to the Socio-Economic Development of Communities: A Case Study of Ming'omba Village, Copperbelt Province. Special Project for Partial Fulfillment of B.sc. Degree in Forestry. School of Natural Resources. Copperbelt University, Kitwe, Zambia.

Njovu, F.C., 1993. Non-wood forest products in Zambia. A country pilot study for the expert consultation for English speaking African countries. In: Commonwealth Science Council and FAO (Eds.). Non-Wood Forest Products: A Regional Expert Consultation for English-Speaking African Countries by FAO. Arusha, Tanzania, 17–22 October 1993.

Nkomesha, A.K., 1997. Baseline Socio-Economic Study of Smallholder Communities in and Around Open Forest Areas and Reserves in Copperbelt Province, Zambia. Provincial Forestry Action Programme (PFAP) Publication No. 20. Forestry Department, Ministry of Environment and Natural Resources, Lusaka, Zambia.

Nswana, A., 1996. Preliminary study on Cosmetic and Traditional Medicine in Central, Copperbelt and Luapula Provinces. PFAP Publication No. 1. Forest Department, Ministry of Environment and Natural Resources. Government of the Republic of Zambia, Lusaka, Zambia.

Packam, J., 1993. The Value of Indigenous Fruit-Bearing Trees in Miombo Woodland Areas of South-Central Africa. Rural Development Forestry Network Paper. <http://www.Odi.org.Uk/Fpeg/Publications/Rdfn/15/Rdfn-15c-Ii.pdf>.

Pegler, D.N., Piearce, G.D., 1980. The edible mushrooms of Zambia. Kew Bull. 35 (3), 475–491.

PFAP, 1998. Socio-Economic Aspects of the Forestry Sector in zambia: An Overview. Forest Department, Ministry of Environment and Natural Resources. Micrographix Express Bureau Ltd., Ndola, Zambia.

Phiri, I.M.G., 2003. An Ecogeographic Study of Selected Vegetatively Propagated Crops in Malawi. SADC Plant Genetic Resources Centre, Lusaka, Zambia, pp. 45.

Piearce, G.D., 1978. Mushrooming in Zambia forestry. Sylva Africana 3, 3.

Piearce, G.D., 1981. An Introduction to Zambia's Wild Mushrooms and How to Use Them. Dunlop Zambia Ltd., Ndola, Zambia, p. 28.

Piearce, G.D., 1982. Nutritive potential of the edible mushroom suillus granulatus (fries) O. Kuntze, and its utilization prospects in Zambia. Trop. Sci. 24 (3), 158—164.

Piearce, G.D., Francis, B.J., 1982. Nutritive potential of the edible mushroom suillus granulatus (fries) O. Kuntze, and its utilization prospects in Zambia. Trop. Sci. 24 (3), 158—164.

PRB Population Reference Bureau 2013. 2013 World Population Data Sheet; Demographic Data and Estimates for the Countries and Regions of the World. Washington, D.C., USA. p. 20.

Puurstjärvi, E., Mickels-Kokwe, M., Chakanga, M., 2005. The Contribution of the Forest Sector to the National Economy and Poverty Reduction in Zambia. Savcor, Lusaka, Zambia.

Roper, J., 1997. Role of Non-Governmental Organization in Forestry Development. Zambia Forest Action Programme (Draft). Forest Department, Ministry of Environment and Natural Resources, Lusaka, Zambia.

Serenje, G., 2002. Micropropagation of Lusala (Dioscorea hirtiflora). A Dissertation Submitted in Partial Fulfillment of Requirements for the Award of the B.Sc. Degree of Agricultural Sciences. Department of Crop Sciences. University of Zambia, Lusaka, Zambia.

Sikombwa, N., Piearce, G.D., 1985. *Vanderbylia ungulata* and its medicinal use in Zambia. Bull. Brit. Mycol. Soc. 19 (2), 124—125.

Storrs, A.E.G., 1982. More About Trees: A Sequel to Know Your Trees. Forest Department, Ndola, Zambia.

Syampungani, S., Meke, G., Geldenhuys, C.J., 2005. Bark wood responses: Results from bark harvesting experiments (Zambia, Malawi and Republic of South Africa). Proceedings of the SADC Regional Workshop for "Trees for Health Forever": Implementing Sustainable Medicinal Bark Use in Southern Africa. Willow Park, Johannesburg, South Africa. 1—3 November 2005.

Taulo, C.M., Mulombwa, J., 1998. The Zambia forestry statistics. FAO Proceedings of Sub-Regional Workshop on Forestry Statistics SADC Region. Mutare, Zimbabwe, 30 November — 4 December, 1998. Forest Department, FAO, Rome Italy.

Vantomme, P., Göhler, D., N'deckere-Ziangba, F., 2004. Contribution of forest insects to food security and forest conservation: the example of caterpillars in the sub-region of the congo basin (Cameroon, Central African Republic, Congo Brazzaville and Democratic Republic of Congo). Wildl. Policy Briefing 3, 1—4.

Vernon, R., 1983. Field Guide to Important Arable Weeds of Zambia. Department of Agriculture, Mount Makulu Central Research Station, Chilanga, Zambia.

Woodward, B., 2000. Natural products, medicinal and ethnobotany: the re-introduction of livingstone potato (P. Esculatus N.E.Br) into a Community. Am. Soc. Plant Biol. 286, 200—205.

CHAPTER 4

Contribution of the Forestry Sector to the National Economy

Phillimon Ng'andwe[a], Jacob Mwitwa[b], Ambayeba Muimba-Kankolongo[b], Nkandu Kabibwa[c] and Litia Simbangala[c]

[a]Department of Biomaterials Science and Technology, School of Natural Resources, Copperbelt University, Kitwe, Zambia; [b]Department of Plant and Environmental Sciences, School of Natural Resources, Copperbelt University, Kitwe, Zambia; [c]Central Statistical Office, Lusaka, Zambia

INTRODUCTION

Several studies on the sector have shown that the current accounting systems, particularly in developing countries, tend to leave out a significant part of the forestry sector contribution to GDP (Gregersen et al., 1997, Poschen, 1997). Consequently, resource allocation to the forestry sector has often been given a low priority because of competing demands from other national developmental sectors and limited government budgets (Puurstjärvi et al., 2005; Ng'andwe et al., 2006). In spite of the many benefits deriving from forests and the great potential that Zambia's forests have in the overall development of the national economy, the achievement of the Millennium Developments Goals (MDGs) and poverty reduction, the country loses between 250,000 and 300,000 ha of forests annually through various destructive activities (Masinja, 2005; MTENR, 2008b). Masinja (2005) considered this as a paradox in that despite abundant forest resources in the country, their contribution to creating wealth and reducing poverty appears almost negligible while at the same time there are very high levels of deforestation. However, recent studies attribute high deforestation rate to the relentless pressure on the forest resources arising from the ever-increasing demand for woodfuel, timber and forestland for other uses such as for agriculture and settlements (Vinya et al., 2012; Ng'andwe, 2012b).

Forest Policy, Economics, and Markets in Zambia. DOI: http://dx.doi.org/10.1016/B978-0-12-804090-4.00004-5

THE ROLE OF THE FORESTRY SECTOR IN THE NATIONAL ECONOMY

It is generally well established that forests provide numerous services that are beneficial to a country's development and livelihood of its population (FAO, 1995; Campbell, 1996; Banda et al., 2008). Livelihoods connote the means, activities, assets and entitlements by which people do make a living. For example, NWFPs collection and their utilization by most communities surrounding the forests as well as their widespread occurrence at both domestic and national markets represent an appropriate reason to enlighten their potential and importance in the daily lives of the population. In addition, forest regulates the flow of water for industry, agriculture and households; protects water catchments from soil erosion, and regulates global climate through maintenance of the balance of atmospheric chemistry. Water that originates from forests provides reliable sources of drinking water, hydroelectricity, industrial development, and irrigation. Furthermore, forests absorb and store carbon dioxide, recycle nutrients, regulate rainfall, and provide the habitat for a diversity of flora and fauna.

As a result of the foregoing there have been many attempts to determine quantitatively as well as qualitatively the real contribution of the forestry sector to the national economy (Makano et al., 1996; Puurstjärvi et al., 2005; Ratnasingam and Ng'andwe, 2012). The contribution of forests to the livelihood security of the population in Africa in general is significant, diversified, and valuable (FAO, 2000; Chademana, 2001). It ranges from direct production of food (Taulo and Mulombwa, 1998) to provision of domestic energy supply, jobs, and supplement of household incomes (Chademana, 2001). Taulo and Mulombwa (1998) and Chademana (2001) give an account of the importance of forests in the socio-economic well-being of Zambians. They found that they provide shelter, food, and medicines, as well as woodfuel necessary for domestic energy. However, the overall contribution of the Forestry sector to the national economy has been grossly undervalued for many years (Chileshe, 2001; Ng'andwe et al., 2006, 2012). Chileshe (2001) reported that this was related to limited statistics available on the forestry-based informal and formal sector activities and also to the classification of wood and wood-based products under the manufacturing industry leading to a high distortion of the true value of the Forestry sector to the national economic development and GDP. Several other reports and frameworks in the past have also highlighted these concerns (Makano et al., 1996; Mulombwa, 1998; Puurstjärvi et al., 2005; Ng'andwe et al., 2008).

PREVIOUS FRAMEWORKS

According to Ng'andwe et al. (2008), considerable efforts have been made to integrate forestry into the national planning process in past years. As early as 1994, the Zambia National Environmental Action Plan (NEAP) of 1994 provided a framework for making significant changes needed to bring environmental considerations into the main stream of decision making. NEAP provided an overview of Zambia's environmental problems, existing legislations, institutions, and various strategic options for improving environmental accounting in the country. Arising from the NEAP was the Zambia Forestry Action Program 2000–2020 (ZFAP) which provides a range of approaches for sustainable forest management (ZFAP, 1997). ZFAP took into consideration the holistic inter-sector and interactive approaches to forest management including ecosystem approaches that integrate the conservation of biodiversity and sustainable use of biological resources, and adequate provision and valuation of forest goods and services. It also included components on wood energy and reforestation. However, forest management has been based on a limited range of economic values often omitting other forest benefits, especially the use of forest products by local communities and environmental services. In 2004, the FAO National Forest Programme Facility (NFP) in partnership with the Forestry Department conducted several studies to:

1. improve the forestry-related information exchange
2. to link forest policy and planning with broader national objectives, strategies and programmes
3. to develop partnerships among stakeholders, and
4. to build the capacity of the Forestry Department staff, NGOs, institutions and the private sector in forest policy development and implementation.

STANDARDS IN COMPUTING ECONOMIC CONTRIBUTION

One of the outputs of the NFP process in Zambia was a study of forest revenue, concession systems and the contribution of the forestry sector to poverty alleviation and Zambia's national economy. The study established that, on average, the forestry sector contributes about 5.2% to GDP per annum (Ng'andwe et al., 2006). Arising from this and other related studies, the preliminary official contribution of the forestry sector to GDP was estimated at 6.3% in 2007 (CSO, 2008) at constant prices of 1994, representing a considerable improvement over earlier estimates of 0.9–3.1% of GDP reported in the years prior to 2006. The misunderstanding was cleared following the domestication and harmonization of certain standards such as the classification and definition of forestry and logging and downstream manufacturing activities under the system of national accounts (UN, 1993) and the use of international standard industry classification (ISIC) of economic activities (UN, 2006a). In addition, other relevant standards such as the National Income Statistics, Sources and Methods (CSO, 1997), Central Product Classification System (UN, 2002), and the Integrated Environmental and Economic Accounting System (UN, 2003) were harmonized through expert stakeholder workshops in 2008.

SYSTEM OF NATIONAL ACCOUNTS

Many countries around the world use the System of National Accounts (SNA) to determine the GDP, a system that was prepared under the auspices of the Commission of the European Union (EU); International Monetary Fund (IMF); Organization for Economic Co-operation and Development (OECD); United Nations (UN) and the World Bank (UN, 2003, 2006a). SNA is a system based on macro-economic accounts intended for use by both national and international statistical agencies, and reinforces the central role of national accounts in economic statistics. Further, SNA provides a coherent, consistent and integrated set of macro-economic accounts, balance sheets and tables based on internationally agreed concepts, definitions, and classifications and accounting rules (UN, 1993). The system also provides an accounting framework within which economic data can be compiled and presented for economic analysis, policy development and decision making processes. The central framework of SNA contains detailed supply and uses tables of records on how supplies of different kinds of goods and services originate, how these are allocated between various intermediates and final users, and tables on records of changes in, and re-evaluation of, assets based on benchmark estimates.

The accounts for goods and services in the SNA trace the economic transactions of products from the original producers to the end-users. The different types of outputs from varied types of production units covering capital goods, intermediate goods, consumption goods; markets, own-account and nonmarket production of goods and services, are all linked and brought together as a single accounting unit. Most national statistics sources and methods in Zambia have been based on the SNA (CSO, 1997). In this system, the aggregate of actual production and final consumption are defined and the compilation of an integrated set of price and volume indices for flows of goods and services as well as the gross value added and GDP are made (CSO, 1997). In the forestry sector of Zambia, the expanded use of SNA is expected to create a satellite framework for the purpose of environmental accounting and to assess interactions between the environment and the economy. It may also be used for delineating transactions of economy within and outside forestry and defining boundaries of products and assets.

INTERNATIONAL STANDARD INDUSTRIAL CLASSIFICATION OF ALL ECONOMIC ACTIVITIES (ISIC)

The ISIC (UN, 2006a) defines an industry as a set of all production units engaged primarily in the same or similar kinds of productive economic activity. Such an activity is characterized by an input of resources, a production process, and an output of products. The ISIC revision 4 classifications consist of sections, divisions, groups, and classes, each of them with a specified scope and coverage of activities (UN, 2006a). Under ISIC, the class of activities represents the activity leading to the production of products classified elsewhere. In this regard, it is harmonized with the commodity description and coding system so that the activity and the resulting product can be related. For example, forest and logging activities are related to the sawn timber produced. However, since it is difficult to meet all the needs for aggregated data, by simple aggregation through various levels of ISIC classification, the user's developed statistics are often a compromise between theoretical principles and practical considerations, making it necessary to domesticate some aspects of the standard within the local classification systems (Ng'andwe et al., 2008).

CENTRAL PRODUCT CLASSIFICATION (CPC)

The Central Product Classification (CPC) constitutes a complete product classification standard, covering transportable and nontransportable goods, services, and assets. CPC includes categories and codes for all products that are transacted domestically and internationally or that can be entered into stocks. It includes not only products that are an output of economic activity (i.e., goods and services from ISIC categories), but also nonproduced assets including tangible assets, such as land, and intangible assets arising from legal contracts, such as patents and copyrights. It takes into consideration the linkage existing between activities and their outputs at a broader level of aggregation (UN, 2002).

In developing CPC, consideration was given to the raw material involved, stage of production, and degree of processing as well as the physical properties, related economic activities, and the purpose or intended use of the product. In this regard, categories of goods in CPC are defined in such a way that each consists of one or more complete Harmonized System (HS) of six-digit category serving as building blocks for the part dealing with transportable goods. In addition, each class of CPC consists of goods or services that are predominantly provided in one class of ISIC. The basic categories of economic supply and the use tables as specified in the SNA have also been taken into account in defining CPC classes (CPC, 2002).

STANDARD INTERNATIONAL TRADE CLASSIFICATION (SITC)

The Standard International Trade Classification (SITC) is a classification made according to physical properties of the product, duly considering the materials from which the product is made and also the stage of fabrication and the industrial origin. The main purposes of SITC are to facilitate international comparison of the product situation, providing greater comparability in foreign trade. This classification system constitutes the basis for a systematic analysis of world trade. The commodity groupings of SITC reflect (a) the materials used in production, (b) the processing stage, (c) the market practices and uses of the product, (d) importance of the commodities in terms of world trade, and (e) technological changes. Since the classification covers only internationally traded commodities, it is dealt with under a five-digit system.

HARMONIZED COMMODITY DESCRIPTION AND CODING SYSTEM (HS)

The Harmonized System (HS) for commodity description and coding is produced by the Customs Co-operation Council system representing the separate categories of goods that corresponds to SITC and the industrial origins of goods (ISIC). It is an international nomenclature for classification of products introduced in 1988 and has been commonly adopted by many countries globally for use in trade statistics.

VALUE ADDED BY THE FORESTRY SECTOR

Data used in national accounts is transformed using ISIC codes and conversion factors and for trade (exports and imports) in accordance with the harmonized system (HS). The contribution from production and trade are usually computed as value added. For simplicity, the Forestry sector contribution is defined as an aggregated value added by forest and logging (including NWFPs and forest service) and manufacturing. The computation of the economic contribution and its linkage to poverty reduction largely depend on the methodology, accuracy of conversion factors and reliability of the baseline data. To compile GDP measures, CSO in Zambia uses various sources of data and methods of estimating value added by each activity and relies on the input from various Government Ministries and the private sector returns to develop tables of conversion factors (CSO, 1997).

Value added is a measure of the contribution to GDP by an individual producer, industry or sector (UN, 2002). The gross value added (GVA) across the sectors is roughly the same as total GDP and is usually calculated at market prices. The gross income or market price is multiplied by the industry average factor for intermediate consumption factor to obtain the GVA. This is the total value of output less the value of intermediate consumption and represents a measure of the contribution to GDP made by an individual producer, industry or sector. GVA may be calculated based on the baseline data or current market prices of products from forest and logging and downstream products under manufacturing following the ISIC and CPC classification systems using the equation 1 shown below:

$$GVA_{\text{forestry sector}} = VA_{\text{forestry and logging}} + VA_{\text{manufacturing wood and wood products}}$$

$$(4.1)$$

The value added (VA) contains the returns to all resource factors employed in the production process, that is, wages and remunerations of employees, profits and surplus margins to resource/capital owners, taxes to government and are estimated as a factor of the gross sales. Computing the value added by a producer is a complex process and is based on a number of assumptions and conversion factors. In general, value added is computed as follows:

$$VA = \text{Remuneration of employee} + \text{Operating surplus} \\ + \text{Government Tax revenue or}$$

$$(4.2)$$

$$VA = \text{Gross sales at market prices} \times f_{VAD}, \qquad (4.3)$$

where:

f_{VAD} is the value added factor for intermediate consumption.

Value added is also a factor of the value of outputs (i.e., a factor to account for intermediate consumption involving remuneration of employees + operating surplus + government tax revenue).

USE OF INTERNATIONAL NOMENCLATURE

The nomenclature derived from CPC, ISIC, and the System of Integrated Environmental and Economic Accounting (SEEA) is a basis for defining the forestry in Zambia for the purpose of determining an inclusive contribution to the national economy (Ng'andwe et al., 2008). The forestry sector's contribution to the national economy is an aggregated value-added contribution from forestry and logging, manufacture of wood and products of wood and manufacture of paper and paper products under the SNA and ISIC.

The ISIC division two (2) includes the production of round-wood and sales to the forest-based manufacturing industries in ISIC division 16 and 17. It also includes the extraction and gathering of wild growing NWFPs for household consumption and trade. In addition, the division includes activities that result in products that undergo little processing, such as collection of firewood, traditional charcoal production, wood chips, and round-wood (cants) used in an unprocessed form (e.g., pit-props, utility poles, etc.). These activities are carried out in natural or planted forests as well as woodlands.

In many cases, when compared with the CPC system, the ISIC represents the economic activity side of the classification such as logging, gathering, etc., while the CPC represents the product side of these two interrelated classifications such as sawn timber, wood-based panels, etc. However, despite the relationships of the two classification standards, the complexity of the relationship between the industry and their products requires careful consideration during the data collection and the analysis stage (CPC, 2002). Hence, some specifications require domestication of the ISIC and CPC systems of classification depending on the use of the information obtained.

USE OF CONVERSIONS FACTORS IN THE FORESTRY SECTOR

The conversion factors of wood and NWFPs is a basis for computing the contribution of the forestry sector to the national economy (Ng'andwe et al., 2012). The International Industry Classification of economic activities (UN, 2006a) is normally used in defining and classifying economic activities of the Forestry sector of Zambia from which conversion factors are derived (Ng'andwe and Njovu, 2007; Ng'andwe et al., 2012; Ng'andwe and Mwitwa, 2012). A conversion factor is defined as using a known figure to determine or estimate an unknown figure via a ratio (UNECE/FAO, 2010). The UNECE/FAO (2010) included the use of the "material balance" which enables a complete understanding of many aspects in the use of forest conversion factors. Generally, the conversion factors are used in analyzing forests and forest products manufacturing efficiencies including determination of silvicultural growth models, biomass calculations, and carbon sequestration in forests, round-wood and wood and wood products, consumption patterns, and export and import transactions. These factors are based on SI units and equivalents to others measures. For example, a saw miller may purchase round-wood in standing cubic meters over bark which has to be converted to under bark volumes and to transport, the saw miller will need to make calculation of the weight to volume ratios.

FOREST AND LOGGING

The forestry sector—forest and logging class includes round-wood, semi-processed cants, fuelwood, charcoal, and NWFPs. Forest and logging conversion factors are based on the rough green of round-wood and physical measurements of outputs.

Round-wood

The common practice of obtaining conversion factors of round-wood involve understanding the physical properties; and using of the method of measuring standing trees and round-wood volume. In this book round-wood volume always means under bark volume and conversion factors are used to convert from one unit of measure to another, for example, from weight to volume. Most sawmills submit data on the conversion factors relating to input and output that is limited to softwood and hardwood. Most operators in timber industry determine true volume using a logical cubic formula and unbiased rounding logic based on different cubic formulas such as Smalian, Huber, Newton, centroid, or two-end conic, (UNECE/FAO, 2010). It is not within the scope of this book to specify a standardized round-wood volume formula suitable for different applications.

Weight and Physical Properties

The weight of round-wood generally correlates well with the volume (UNECE/FAO, 2010). In Zambia round-wood is bought by volume and transported via weight. Usually weight is used in conjunction with sample volume measurement in order to establish the relationship. The presence of drive-on weight scales along some transportation routes make weight data readily available and inexpensive to ascertain relative to taking measurements on all logs. There are a number of factors such as basic density, moisture content, and bark, etc., which determine the weight of a given amount of round-wood volume. It is assumed that approximately 66.7% of the displaced volume is wood, 11.5% is bark, and 21.6% is void space within a stack (UNECE/FAO, 2010).

Wood Basic Density

Conversion factors for wood density is typically measured as a ratio of the weight of oven dry wood and volume in kilograms per cubic meter. It is also measured as an index of the relationship of wood to the same volume of water (water weighs 1,000 kg/m^3). For example, a cubic meter of pine softwood wood without any water and weighing 420 kg has a basic density of 420 kg/m^3 and a specific gravity of 0.42. In Zambia basic density (green volume and oven dry weight) has been estimated to vary from approximately 420 kg/m^3 to 540 kg/m^3 for softwoods, while hardwoods vary from about 650 kg/m^3 to 850 kg/m^3 and shrinkage is estimated at 10% and 15% for softwood and hardwoods, respectively (UNECE/FAO, 2010; Ng'andwe and Mwitwa, 2012). The exact amount of shrinkage varies from one species to another. However, basic density based on green volume may be used when computing standing tree volume and round-wood without having to know or estimate volumetric shrinkage (UNECE/FAO, 1987).

Wood Moisture Content

Wood when freshly cut contain contains water in lumens, cell cavities, and within the cell walls. Moisture content is measured in terms of the weight of the moisture relative to the oven dry weight of wood. Freshly cut timber can vary from 30% moisture content dry (i.e., mc data at fiber saturation point) to more than 200%. It is important to take into consideration the variations between species, season of softwood and hardwoods, heartwood, and sapwood, as well as age of a tree. It is assumed that older trees often have a lower weight to volume ratio than that of younger trees by virtue of age related increase in the ratio of heartwood to sapwood.

Utility Poles

Conversion factors here are reported in numbers of poles, volume, and weight. The average sizes range from 15–30 cm (top) × 6–12 m long for industrial poles (ZAFFICO, 2007). At household level, sizes are in the range of 5–10 cm top diameter and this is usually reported in numbers which is computed by using an average pole size of 8 cm top diameter and length of 10 m. The average volume used for a typical utility pole is 0.05 m^3 for indigenous poles and 0.06 m^3 for plantation poles, while the conversion factor is estimated at 1.42 m^3 of round-wood input per cubic meter of finished utility pole.

Conversion for Estimating the Value Added

A typical example of use of conversion factors for estimating the value added by each ISIC group is based on volume values as illustrated in Box 4.1.

In SNA, the boundary between forest and logging (section A) and manufacturing (section C) is not crossed when computing the contribution of each economic activity. The recent trend, however, is to aggregate these sections for the purpose of determining an aggregate and inclusive contribution of the sector to the national economy. The gross value added is, therefore, the total value of output less the value of intermediate consumption and is a measure of the contribution to GDP made by an individual producer, industry, or sector. For the forestry sector of Zambia, this is an aggregated value-added by forestry and logging and manufacturing (Box 4.1).

Various studies in recent years (Puurstjärvi et al., 2005; Ng'andwe et al., 2006, 2008) have used estimated conversion factors to compute the contribution of the forestry sector to the national economy and poverty reduction. Puurstjarvi (2005) and Ng'andwe et al. (2006) observed that the contribution of the forestry sector to the national economy and poverty reduction has usually been under estimated in

Box 4.1 Factors for Computing the Contribution of the Forestry Sector to the National Economy in Relation to International Standard Industry Classification of Economic Activities

Section	Division	Group	Class	Activity Description	WRME (m³ rw/m³ p)	Factors Based on Market Price or Output Value	
						Intermediate Consumption	Value Added
A				**Agriculture, Forestry, and Fishing**			
	1			Crop and animal production			
	2				**Forestry and logging**		
		21	210	Silvicultural activities		0.20	0.80
		22	220	Logging	1.25	0.30	0.70
			221	Round-wood production	1.77	0.35	0.65
			222	Utility poles production	1.77	0.35	0.65
			223	Gathering of firewood	1.33	0.25	0.75
			224	Charcoal production	5.26	0.35	0.65
		23	230	Gathering of NWFPs	N/A	0.75	0.35
		24	240	Support services	N/A		
			241	Forest inventories	N/A		
C	16			**Manufacture of wood and products of wood**			
		161	1610	Sawmilling	2.45	0.70	0.30
		162	1621	Veneer production	2.86	0.50	0.50
			1622	Builders carpentry	3.77	0.30	0.70
			1623	Wooden containers	3.77	0.30	0.70
			1624	Other products	3.77	0.50	0.50
	17	170	1700	Paper and paper production		0.5	0.50

Note: *WRME is the wood raw material equivalent needed to produce one cubic meter of output.*

Table 4.1 Conversion Factors for Round-wood Production Wood Material Balance Under Forest and Logging

Forest and Logging	Unit Output	Size of Enterprise			Average
		Large	Medium	Small	
Softwood					
Round-wood green	m³ rw/m³ p	1.18	1.43	1.67	1.42
Material balance		100	100	100	100
− Round-wood	%	85	70	60	71.7
− Poles	%	0	0	0	0.0
− Residues	%	10	20	25	18
− Fuelwood	%	0	5	10	5
− Wood sawdust	%	5	5	5	5
Hardwood					
Round-wood green	m³ rw/m³ p	2.00	1.82	2.50	2.11
Material balance		100	100	100	100
− Round-wood	%	85	70	60	71.7
− Poles	%	0	0	0	0.0
− Residues	%	10	20	25	18
− Fuelwood	%	0	5	10	5
− Wood sawdust	%	5	5	5	5

Note: m³ rw/m³ p means cubic meter round-wood required to produce one cubic of product.

many computations. In many cases, such attempts have used different methods, conversion factors, definitions, and standards other than those used by CSO in the national accounts (CSO, 1997; UN, 2006b; Ng'andwe et al., 2008). In the forestry sector, the starting point in determining conversion factors in the downstream industries is the forest and logging class. Forest and logging class includes round-wood and poles, semi-processed sawn wood, fuelwood, and charcoal as main economic activities selected for the purpose of determining conversion factors (Table 4.1 and 4.2).

CONVERSION FACTORS FOR ENERGY PRODUCTION FROM WOOD

Wood biomass is often used to generate energy and the conversion factors are also applied. For example, if there is need to set up a bioenergy plant from forest and mill biomass to generate electricity it is important to know how much energy can be produced apart from

Table 4.2 Conversion Factors for Woodfuel Production Material Balance Under Forest and Logging

Description act	Unit Input per Unit Output	Comparison	
		Zambia	UNECE
Fuelwood	**m³ rw/odmt**	1.33	2.1
Product basic density (solid volume, oven dry)	kg/m³	650	521
Higher heating value	m³ rw/gj	–	0.12
Pellets	**m³ rw/m³ p solid**	–	2.54
Round-wood input to bulk m³ pellets	m³ rw/m³ p bulk	–	1.52
Product basic density (solid volume, oven dry)	kg/m³	–	1067
Bulk basic density (solid volume, 5–10% mcw)	kg/m³	–	677
Higher heating value (bulk volume)	m³ bulk/gj	–	0.09
Briquettes	**m³ rw/odmt**	–	**1.96**
Product basic density (solid volume, oven dry)	kg/m³	–	1075
Bulk basic density (solid volume, 5–10% mcw)	kg/m³	–	761
Higher heating value	m³ rw/gj	–	0.09
Bark and chipped fuel	**m³ rw/odmt**	–	**2.39**
Product basic density (solid volume, oven dry)	kg/m³	–	373
Bulk basic density (solid volume 5–10% mcw)	kg/m³	–	2.36
Higher heating value	m³ rw/gj	–	0.1
Charcoal	**m³ rw/odmt**	5.26	5.96

Note: m^3 rw/m^3 p means cubic meter round-wood required to produce one cubic of the product, m^3 rw/odmt means cubic round-wood required to produce one oven dry metric tone, m^3 rw/gj means cubic round-wood required to produce one giga joule of energy.

knowledge of calorific values. In general or as rule of thumb, 1 ton of dry wood can generate 5 MWh, and can generate about 0.25 MWh of electricity as illustrate in Box 4.2.

If the energy is being transmitted or used at a constant rate (power) over a period of time, the total energy in kilowatt-hours is the product of the power in kilowatts and the time in hours. In order to know how much forest and biomass is required to feed the biomass power plant this basic conversions may be used as a rule of thumb during prefeasibility studies. For example, to run a 10 MW biomass plant will require a supply of 120,000 tons of wet wood or 350,000 MWh from the forest and mill biomass. When electricity is sold, kilowatts are used as a billing unit for energy delivered by the power supplier to the consumer. The kilowatt-hour (symbolized kWh) is a unit of energy equivalent to one kilowatt (1 kW) of power expended for one hour. Since one watt is equal to 1 Joule per second (J/s), one kilowatt-hour is 3.6 megajoules, which is the amount

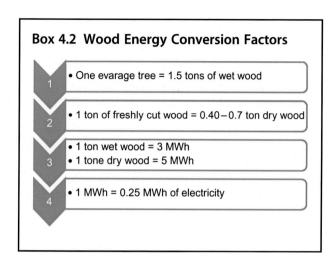

Box 4.2 Wood Energy Conversion Factors

1
- One evarage tree = 1.5 tons of wet wood

2
- 1 ton of freshly cut wood = 0.40–0.7 ton dry wood

3
- 1 ton wet wood = 3 MWh
- 1 tone dry wood = 5 MWh

4
- 1 MWh = 0.25 MWh of electricity

of energy converted if work is done at an average rate of one thousand watts for one hour. When wood is used to produce energy other basic conversion factors may be used such as tons of oil equivalent (TOE).

One TOE is equivalent to 42 GJ (net calorific value). It is estimated that one ton of fuelwood is equivalent to 0.3215 TOE. Therefore it is possible to use this factor to calculate TOE from different energy sources within the national energy mix. For example, in 1996 the energy consumed in Zambia exceeded 4.5 million TOE per annum (i.e., Wood fuel 68%, Petroleum 14%, Electricity 12%, and Coal 6%). Ng'andwe et al. (2006) computed a total of 498,800 TOE from the 5.8 GWh of electricity consumed in 2003 and based on this conversions, they concluded that the value-added by wood energy (i.e., USD374 million) was much higher than the value of electricity generated. The value added by wood energy can be invested into hydro-electricity which has always been an important part of the world's electricity supply, providing cost-effective energy. It is recognized that much of the remaining hydro potential in the world exists in the developing countries of Africa and Asia. While these alternatives are being considered there, must be deliberate macroeconomic policies that are directed to start managing forests sustainably.

CONVERSION FACTORS UNDER MANUFACTURING

The manufacturing class of ISIC includes green sawn wood, veneer, plywood and particleboard main economic activities selected for the purpose of determining conversion factors and wood raw material balance (Tables 4.3, 4.4, and 4.5).

Table 4.3 Conversion Factors and Wood Raw Material Balance for Sawmill Industry						
Description	Unit Input per Unit Output	Size of Enterprise			Zambia	UNECE
		Large	Medium	Small		
Softwood						
Sawn wood green	m^3 rw/m^3 p	2.00	2.50	2.86	2.45	1.68
Cants green	m^3 rw/m^3 p	1.25	1.43	2.00	1.56	1.49
Hardwood						
Sawn wood green	m^3 rw/m^3 p	2.86	3.57	4.00	3.48	1.83
Cants green	m^3 rw/m^3 p	1.33	1.43	2.22	1.66	1.37
Material balance softwood		100	100	100	100	100
− Sawn timber	%	45	35	30	36.7	54
− Chips/slabs	%	34	22	32	29.3	29
− Shavings	%	8	10	0	6	2
− Wood sawdust	%	5	25	30	20	11
− Shrinkage loss	%	8	8	8	8	5

Table 4.4 Conversion Factors and Wood Raw Material Balance for Plywood Industry					
Description	Unit Input per Unit Output	Size of Enterprise		Zambia	UNECE
		Large	Medium		
Hardwood and softwood					
Rotary cut veneer, dry	m^3 rw/m^3 p	2.22	-	2.22	1.97
Sliced veneer, dry	m^3 rw/m^3 p	2.86	-	2.86	2.77
Plywood dry sanded	m^3 rw/m^3 p	4.00	-	4.00	2.12
Material balance	%	100	-	100	100
− Veneer	%	25	-	25	49
− Others residues	%	65	-	65	42
− Sanding	%	5		5	4
− Shrinkage/losses	%	5	-	5	5
Average shipping weight	Kg	650	-	650	584
Average panel thickness	Mm	16		16	16.5

CONTRIBUTION TO THE NATIONAL ECONOMY

It has been estimated that the Forestry sector contributes an average of 6.2% to the GDP every year (CSO, 2010; Ng'andwe et al., 2012). The forest and logging has the largest share (79%), followed by wood and wood products (14%), and paper and paper products with a 7%

Table 4.5 Conversion Factors and Material Balance for Particleboard					
Description	Unit Input per Unit Output	Size of Enterprise		Zambia	UNECE
		Large	Medium		
Particleboard	m³ rw/m³ p¹	1.54	1.67	1.60	1.51
Average thickness	mm	16	16	16.00	18.3
Product basic density	kg/m³	650	650	650	661
Material balance		100	100	100	100
− Binders and fillers	%	4	4	4	4
− Bark	%	2	2	2	0
− Moisture	%	10	10	10	7
− Wood	%	84	84	84	90

Note: m³ rw/m³ p means cubic meter round-wood required to produce one cubic of the product.

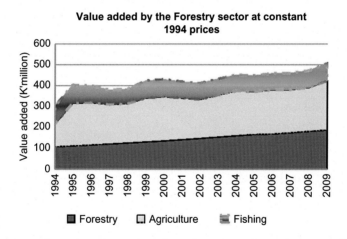

Figure 4.1 The aggregated value added contribution by ISIC division A: Agriculture, Forestry and Fishing.

contribution (Ng'andwe et al. (2012)). In 2010, the share of the forestry sector to GDP was 5.7%, with forestry and logging contributing 4.5%, while wood and wood products made up to 1.2%.

The Forestry sector Gross Value Added statistics are normally obtained at constant market prices of a baseline year. The value added contribution by ISIC division A (i.e., agriculture, forestry and fishing) has been increasing over the years from K300 million in 1994 to K506 million in 2009 (Figure 4.1). Within this division, agriculture has the largest gross value added followed by forestry and then fishing.

Figure 4.2 Value added trends since 1994 by the Forestry sector.

Table 4.6 Industry % Shares of GDP by Kind of Economic Activity at Constant 1994 Prices						
Kind of Economic Activity	Annual Share of GDP (%)					
	2005	2006	2007	2008	2009	2010*
Round-wood production	5.2	5.0	4.9	4.8	4.7	4.5
Wood and wood products	0.8	0.8	0.8	0.8	0.8	0.8
Paper and Paper products	0.3	0.3	0.3	0.3	0.3	0.4
Total Forestry sector	*6.4*	*6.1*	*5.9*	*5.9*	*5.8*	*5.7*
TOTAL GDP AT MARKET PRICES	100	100	100	100	100	100
Note: GDP means gross domestic product.						

Wood and wood products include sawn timber and manufacture of wood-based products. There has been an increasing trend in the value added contribution from manufacturing to the Forestry sector gross value added total from ZMK130 million in 1994 to ZMK225 million in 2009 (Figure 4.2). Gross Value Added (GVA) is a measure of the contribution to Gross Domestic Product (GDP) by an individual producer, industry or sector (UN, 2002). The activities by wood manufacturing include round-wood production, wood, and wood products, as well as paper and paper products.

In spite of the stable level of valued added in the forestry sector over the years, its contribution to GDP has declined from 6.4% in 2005 to 5.7% in 2010 (Table 4.6). The decline could be attributed to a combination of factors including loss of vegetation cover as a result of deforestation and degradation, the global financial crisis experienced by World economies, etc.

The higher contribution to GDP from primary activities such as forestry supplying raw materials such as round-wood through licensing than forest products processing indicates the lack of value-addition in the downstream industries. In the vision 2030 (GRZ, 2006) the Government of Zambia outlines the plans to increase the contribution to GDP from secondary sectors (e.g., wood and wood products) to be higher than round-wood production, licensing and environmental services to address the need for value addition.

TRENDS OF THE SECTOR GROWTH RATES

Zambia's industrial output increased by 5.6% in 2010 and the growth was attributed to manufacturing and electricity, which grew by 10.7% and 10.1%, respectively (CSO, 2012). On the other hand, the manufacturing industry experienced growth in food, beverages and tobacco; nonmetallic mineral products; paper and paper products; and metal by-products. There was, however, a decline in the overall growth of wood and wood products from 13.4% in 2010 to 6.1% in 2011 and that of paper and paper products from 22.7% to 18.4%, respectively, due to the overall poor economic performance arising from factors mentioned earlier in the section.

According to the vision 2030 of Zambia (GRZ, 2006) and SNDP (GRZ, 2011), the contribution of agriculture, forestry and fishing sector to GDP is expected to decline from 23.6% in 2004 to 10.1% in 2030 and that of mining from 3.4% in 2004 to 2.4% in 2030 in line with the expected increase in value added processing in manufacturing. Innovation from private corporations and the provision of enabling environment for production, processing, and delivery of value added goods would be key strategies in achieving and maintaining such milestones. In the manufacturing of wood and wood products such strategies could include provision of incentives for value-added processing, establishment of kiln-drying, and grading of timber. Additionally, the development of quality wood furniture facilities for MSMEs clusters especially in the Copperbelt and Lusaka Provinces which have comparative advantages over other Provinces in relation to the existing of industrial infrastructure will be an added milestone (SNDP, 2011). To comply with this, it is important to invest in appropriate, cleaner and environmentally friendly technologies, create linkages with potential investors, develop incubation programs to upgrade small and medium-sized producers which could embark on finished product processing, and train MSMEs in appropriate skills such as timber grading and entrepreneurship (Ratnasingam and Ng'andwe, 2012).

STATUS AND CHARACTERISTICS OF THE FORESTRY SECTOR

Strengths, Weaknesses, Opportunities, and Threats

Although Zambia has a high forest resource per capita (3.76 ha/ person) among other countries in the region, the exploitation of the forest resources is skewed toward low value-added commodity products such as production of cants, rough sawn timber and untreated poles (Ratnasingam, 2012). Because it is a land-locked country, Zambia is constrained by the limited accessibility to international markets through transportation by sea. Because of its high cost, land transportation adversely affects the pricing of various products at marketplaces. In order to evaluate the potential of the forestry sector in the country, Strengths, Weaknesses, Opportunities, and Threats (SWOT) may be used as a tool (Ratnasingam and Ng'andwe, 2012). The SWOT analysis and evaluation of the forest products industry processing could reveal the future potential for expanding the value-added wood products industry. Such an in-depth industry analysis could result in calling for a paradigm reform shift both in terms of resource management, industrial planning, and adjustments in the characteristics of the wood sector. The SWOT summary of the Zambia Forestry sector provides a starting point for industry adjustments (Table 4.7).

The industry of forest products in the country has a number of opportunities for enhancing marketing of wood products, particularly as contained in the Government vision 2030 and sixth national development plans. Through these plans, the Government aspires to promote the establishment of a profitable environment for increased domestic industrial growth, housing and export activities, the development of market-oriented production management, and the scaling-up of the private sector (SADC, 2006; GRZ, 2006, 2011). Zambia suffers from a serious housing shortage that is worsening as the population grows. Ratnasingam and Ng'andwe (2012) predicted that the housing

Table 4.7 The SWOT Summary of the Forestry Sector in Zambia	
Description	Result
STRENGTHS	– large wood resource – large workforce at low industry wages (USD 200/month) – centrally positioned in Southern Africa, which makes neighboring countries potential markets – previous inherent experience in forest planning and management practices
WEAKNESSES	– industrial production dependent on plantation resource, which is dwindling – weak institutions in terms of forest management, implementation and enforcement, that is, too much "rent seeking" – business development services, such as quality enhancement, licensing and provision of know-how are grossly inadequate – prevalent low-wage economy deters trained workers from entering the wood industry – insufficient or absent supporting industries, such as kiln-drying facilities, machinery suppliers, tooling, etc., makes value-adding difficult – high transportation and energy cost – extensive use of out-dated technologies – capital availability is limited/cost of financing is too high
OPPORTUNITIES	– establish a formal "Timber Industry Development Agency" that is empowered to incentivize and manage the timber industry – bring all other organizations as well as trade bodies into this umbrella organization to voice out and coordinate the needs of the industry – hold a national wood products exhibition to showcase and promote products and also allied services, that is, education, training, quality, investment opportunities, etc. – formalize the market structure with clear distribution channels to ensure quality and value-adding activities – targeted incentives to boost value-adding must be provided – institute a legal framework to regulate the industry, both in terms of planning and management
THREATS	– the cost of wood fiber in the country is relatively low, which explains the high waste at the various level of processing – large mills or foreign investors have a free hand to exploit the resources to the maximum, depriving the local communities of the legitimate opportunity – the accelerated exploitation of selective valuable timbers will deplete the value of the forest as a whole

and construction sector will remain the largest consumer of wood and wood products in the country in years to come. However, the domestic market is predominantly dominated with untreated sawn timber used for construction, builders joinery and carpentry and furniture, etc. It also includes transmission poles treated with chemicals such as copper chromate arsenic (CCA) and creosote. In order to boost exports of wood and wood products there is a need to :

• adhere strictly to quality codes and export procedures to ensure that value addition takes place;
• enforce and legislate frameworks for wood resource exploitation and processing;

- introduce grading rules, kiln-drying and treatment compliances;
- formulate incentives to boost recovery of waste and other mill residues.

Constraints to the Development of the Market

There are many constraints that hinder the potential development of the wood product industry in Zambia, and among the most important are:

1. The uneconomically low prices of sawn timber and wood products due to quality issues arising from the lack of quality assurance standards and absence of a designated market (physical or otherwise) which is largely controlled by timber traders and other wood producers.
2. The lack of wood kiln-drying capacity and treatment facilities reduce the shelf life of timber before it becomes dis-colored, bent, warped or cracked as it lies in the stacks at the saw-mills or at trading points. The wood products industry would need to significantly increase its kiln-drying and treatment capacities.
3. Local market demand for 5 m length products results in major losses of roundwood in the forest plantations of more than 20% of purchased volume as no proper scaling is done due to such market requirements. Inevitably, it leads to over-harvesting at the plantations which places considerable pressure on the growing plantation stock of tree species like pine and eucalyptus.
4. Disposal of sawmill 'waste' such as sawdust poses a huge challenge to the sawmillers and the local authorities mainly as the mountains of sawdust found all over towns have become environmentally hazardous. Further, the disposal of sawdust constitutes also a significant cost for the micro and small sawmill operations.
5. Lack of affordable finance to ensure smooth running of operations, organize training for foresters and sawmillers and invest in new and efficient machinery and log transport vehicles.

In general, the forestry and wood products industry is affected by numerous factors including external (policy and regulatory frameworks as well as those specific to industries) and intrinsic factors related to general industrial operations.

LINKING THE FORESTRY SECTOR CONTRIBUTION
TO THE UN-REDD

It is stipulated in the land ownership structure in Zambia that communities own and control over 45,000,000 ha (61%) of land out of which 35,300,000 ha (78%) are forests and woodlands under government control (MTENR, 2008a). The private timber processing companies are given concession licenses which are obtained with consent from community's first. The legislation requires that communities get involved in the negotiation for the use rights and responsibility regarding the forest at the earliest stage of the licensing process. However, it is evident from the forest classification in relation to the potential yield that a larger proportion of the forests are also used to collect wood and NWFPs by the communities for livelihood (Ng'andwe et al., 2006; Ng'andwe, 2012a). These activities by communities and industry have been considered to be among the major drivers of deforestation in the country (Vinya et al., 2012). Therefore, there is a need for all actors to be sufficiently rewarded if change in the way communities and industry use forests to reduce deforestation and forest degradation is expected.

Ratnasingam and Ng'andwe (2012) reported that afforestation and reforestation programs constitute a good source of emission reduction through the Clean Development Mechanism (CDM) and the voluntary carbon markets. This implies that with good structures in place, REDD+ benefits can be transferred to smallholder land owners and communities following a set of eligibility rules that determine under which scheme the activities can be implemented for carbon benefits. Since REDD+ requires that baseline data for biomass (in volume or tons) and carbon stocks (tons) for forest types are determined, inventory of forest biomass and merchantable timber should be determined in the country using acceptable parameters and conversion factors that have been calculated. Chidumayo (2012) proposed interim biomass combustion efficiencies (CEs) and emission factors (EFs) reference levels for Zambia based on the 1965–2005 data which should be considered in the integrated land use planning. In addition, reforestation for renewable biomass production as a forest opportunity potential under the CDM can only be done to land that was not forested prior to 1990 (Smith, 2002; Mwanka, 2012). This is an opportunity for CDM projects in forests which were used under coup systems for charcoal production before 1990 but that which can be managed for carbon markets by communities and the private sector in order to increase the share of the value added into the national economy.

It is therefore imperative to take action in the planning process for forest management and processing to determine the extent to which timber processing companies contribute to deforestation and institute necessary measures to reduce emissions. In line with this, Chidumayo (2012) indicated that the first approach would be to map out future deforestation hot-spots based on human population growth projections for which time-series data are available including those areas that have been under concessions and coup systems before 1989. Afforestation and reforestation of degraded forest areas in the hot spots can be integrated in this approach. Therefore, Zambia forestry and forest industries, Forestry Department and other private companies have the opportunity to partner and embark on a country-wide afforestation and reforestation program through public-private partnership and be able to access benefits from REDD initiatives (GRZ, 2006, 2010, 2011) and from CDM. This opportunity should also be explored by taking advantages of and adopting Nationally Appropriate Mitigation Actions (NAMAs) that funds arrangements that potentially cover all sectors.

CONCLUSION

There is need for increased participation by the private sector in the planning and development of sustainable wood supply in the country by preventing illegal logging, formulating of regulations and enabling policies that will result in the growth of the industry. Strategies must include measures such as provision of incentives for value added processing and establishment of kiln-drying and wood furniture facilities for MSMEs clusters especially in the Copperbelt and Lusaka Provinces. In addition, forestry industrial development must be encouraged and helped to start investing in appropriate, cleaner and environmentally friendly technologies by linking them with potential investors, developing incubation programs to upgrade small and medium-sized producers to embark on value added product processing, and providing training in appropriate skills such as timber grading and entrepreneurship.

The success of the Zambian wood products sector to move forwards to greater value-added activities would reside on its ability to adopt a new approach with a remarkable paradigm shift. This calls for techno-logical, skill, knowledge, policy, and regulatory improvements that would lay the platform for such industrial transformation. It also calls for strengthening of related institutions, particularly the forest associa-tions and the Forestry Department to enable better planning, manage-ment, and control of the wood products industry. The possible benefits that could derive from the CDM, UN-REDD, and FLEGT schemes need to be exploited to ensure that all stakeholders, particularly the local communities, embrace and benefit from the principles of these schemes.

In order to realize the full potential of the contribtion of the forestry sector to the national economy the following specific recommendations have been endorsed at various fora of the experts in the sector in Zambia:

Government should establish new Fiscal Instruments that:

- Provide incentives for value adding, especially handling of waste as a raw material
- Provide incentives to employ and retain workers
- Provide incentives for applied research into lesser known tree species and lesser used species in collaboration with the Forestry Department and higher learning institutions in forestry
- Attract investors to deal with waste utilization (finger-jointing/ wrapping of solid wood, pellets and briquettes, etc.) by providing adequate incentives such as tax breaks, etc.

Human capital development:

- Start apprentice-scheme for school leavers
- Create markets or government-linked cooperatives to buy back forest products from students in order to encourage entrepreneurship
- Upgrade facilities in training institutions
- Tax breaks for industry that accepts students for industrial training
- Curriculum revision

A "Social Contract" for the wood industry:

- Engage community through cluster development systems
- Forestry sector and the Zambia Development Agency (ZDA) should explore skills enhancement training for these communities in order to participate in low risk products development (e.g., school for fur-niture manufacturing)
- Establish tree plots for each community to create forest plantations, etc.

Technology upgrade:

- Create a dictionary of Zambian timbers for all commercial species and lesser known species
- Make information available and accessible and disseminate actively the information
- Upgrade technology in all sub-sectors dealing with wood and also enhance workers' skills—which will increase wage and value adding, while reducing waste
- ZDA should take relevant players to machinery exhibition, etc.

Economic foundation:

- Forestry Department needs enhance its capacity in order to increase revenue collection
- Develop a Master Plan for the Timber Industry for the next 10 years
- Make the timber industry effectively empower the rural community

Markets:

- Enforce compulsory requirements for kiln-drying and wood treatment
- Remove distorted market structures—by increasing stumpage price or price per tree currently in schedule of payment
- Specification and verification of wood products must be enforced—by establishing a quality certificate for "Zambian Wood Products" to be issued by the Forestry Department or other related agencies
- Establish the Timber Industry Development Agency to plan and manage the wood products sector

REFERENCES

Banda, M.K., Ng'andwe, P., Shakacite, O., Mwitwa, J., Tembo, J. C., 2008. Markets for Wood and Non-Wood Forest Products in Zambia. Final Report Submitted to FAO-NFP.

Campbell, B., 1996. The Miombo in Transition. Woodlands and Welfare in Africa. CIFOR, Bogota, Indonesia.

Chademana, S.N., 2001. Improving Food Security in Zambia: The Ignored Role of Forestry. Special Project for B.Sc. Degree in Forestry. School of Natural Resources. Barchelor of Science, Copperbelt University, Kitwe, Zambia.

Chileshe, A. 2001. Forestry Outlook Studies in Africa (FOSA) — Ministry of Natural Resources and Tourism — Zambia. Available online: <http://www.fao.org/forestry/FON/FONS/outlook/Africa/AFRhome>.

CSO, 1997. National Income and Statistics Sources and Methods, vol. 1. Bench Mark Estimates 1994, National Accounts. Central Statistical Office, Lusaka, Zambia.

CSO, 2008. Central Statistical Office. Gross Domestic Product 2005 Revised Estimates Real. Ministry of Finance and National Planning, Lusaka, Zambia.

CSO, 2012. CSO monthly bulletin, <www.zamstats.gov.zm/media.php>.

FAO, 1995. Valuing Forests: Context, Issues and Guidelines. FAO Forestry Paper No. 127. Department of Forestry, Rome, Italy.

FAO, 2000. Statistical Data on Non-Wood Forest Products in Africa: Draft report. EC-FAO Partnership Programme. Forest Department, FAO, Rome, Italy.

Gregersen, H., Lundgren, A., Kengen, S., Byron, N., 1997. Measuring and Capturing Forest Values: Issues for the Decision-Maker. Paper Prepared for the Eleventh World Forestry Congress, 13–22 October 1997, Antalya, Turkey.

GRZ, 2006. In: GRZ (Ed.), Government of the Republic of Zambia Vision 2030. Government Printers, Lusaka, Zambia.

GRZ, 2010. UN Collabnorative Programm on Reducing Emmissions from Deforestation and Forest Degradation in Developing Countries National Joint Programme Document — Zambia Quick Start Initiative. In: MTENR (Ed.). Lusaka, Zambia.

GRZ, 2011. Government of the Republic of Zambia Sixth National Development Plan. Ministry of Finance and National Planning. Government Printers, Lusaka, Zambia.

Makano, A., Ngenda, G., Njovu, F., 1996. The Contribution of Forestry Sector to the National Economy. A Task-Force Report for ZFAP. Department of Forestry, Ministry of Environment and Natural resources, Lusaka, Zambia.

Masinja, A., 2005. Keynote Address. In: The Proceedings of the First National Symposium on the Timber Industry in Zambia 29–30 September. Mission Press, Ndola, Zambia. In: Ng'andwe, P. (Ed.), National Timber Symposium. Mulungushi Conference Centre, Lusaka, Zambia.

MTENR, 2008a. Ministry of Environment and Natural Resources Management and Mainstreaming Programme. Lusaka, Zambia.

MTENR, 2008b. Integrated Land Use Assessment 2005–2008. Forestry Department, Ministry of Tourism, Environment and Natural Resources, Lusaka, Zambia.

Mulombwa, J., 1998. Non-Wood Forest Products in Zambia. A Report of the EC-FAO Partnership Programme (1998–2000). Project GCP/INT/679/EC Data Collection and Analysis for Sustainable Management in ACP Countries — Linking. National and International Efforts. Forest Department, FAO, Rome, Italy.

Mwanka, R. D., 2012. Guidelines for Clean Development Mechanism and Voluntary Carbon Projects in Zambia. Report Submitted to FAO National Forest Programme (NFP) Facility and Forestry Department. MTENR, Lusaka, Zambia.

Ng'andwe, P., 2012a. Forest Classification, Zones and Classes — Basis for Industrial Processing. Submitted to the Ministry of Lands, Natural Resources and Environmental Protection and FAO, Lusaka, Zambia.

Ng'andwe, P., 2012b. Forest Products Industries Development — A Review of Wood and Wood Products in Zambia. Submitted to the Ministry of Lands, Natural Resources and Environmental Protection and FAO, Lusaka, Zambia.

Ng'andwe, P., Mwitwa, J., 2012. Conversion Factors for the Forestry Sector Statistics Under the National Forest Program Facility, FAO. In: Ng'andwe, P., Mwitwa, J. (Eds.), Conversion Factors for the Compendium of Forestry Statistics. School of Natural Resources: Copperbelt University, Kitwe, Zambia.

Ng'andwe, P., Njovu, F., 2007. Round Wood Supply and Demand Trends in Zambia and the SADC Region, Report submitted to Zambia Forestry and Forest Industries Corporation. Kitwe, Zambia.

Ng'andwe, P., Muimba-Kankolongo, A., Shakacite, O., Mwitwa, J., 2006. Forest Revenue, Concessions Systems and the Contribution of the Forestry Sector to Zambia's National Economy and Poverty Reduction. FEVCO, Lusaka, Zambia.

Ng'andwe, P., Muimba-Kankolongo, A., Mwitwa, J., Kabibwa, N., Simbangala, L., Mulenga, F., 2008. Forestry Sector Guidelines for Data Collection and Handling. Forestry Guidelines, Lusaka, Zambia.

Ng'andwe, P., Simbangala, L., Kabibwa, N., Mutemwa, J. 2012. Zambia Biennial Compendium of Forestry Sector Statistics 1980−2010. Compendium, Ndola, Zambia.

Poschen, P., 1997. Forests and Employment much more than Meets the Eye. Paper Prepared for the Eleventh World Forestry Congress, 13−22 October 1997, Antalya, Turkey.

Puurstjärvi, E., Mickels-Kokwe, M, Chakanga, M., 2005. The Contribution of the Forest Sector to the National Economy and Poverty Reduction in Zambia. SAVCOR, Lusaka, Zambia.

Ratnasingam, J., 2012. Estimated Future Potential for Expanding Value Wood Processing Activities based on Zambian Wood. Submitted to the Ministry of Lands, Natural Resources and Environmental Protection and FAO, Lusaka, Zambia.

Ratnasingam, J., Ng'andwe, P., 2012. Forest Industries Opportunity Study — Synthesis Report Submitted to the Forestry Department Integrated Land Use Assessment II and the Food and Agriculture Organisation (FAO) of the United Nations, Lusaka, Zambia.

SADC, 2006. Trade, Industry and Investment Review. <http://www.sadcreview.com/country_profiles/zambia/zambia.htm>.

Smith, J., 2002. Afforestation and reforestation in the clean development mechanism of the Kyoto protocol: implications for forests and forest people. Int. J. Global Environ. 2, 322−343.

Taulo, C. M., Mulombwa, J., 1998. The Zambia Forestry statistics. FAO Proceedings of Sub-Regional Workshop on Forestry Statistics SADC Region. Mutare, Zimbabwe, 30 November — 4 December, 1998. Forest department, FAO, Rome Italy.

UN, 1993. System of National Accounts. Prepared under the Auspices of Eurostat, IMF, OECD, United Nations and World Bank. United Nations, Dept. of Economic and Social Affairs. Statistical Office New York, Brussels, Washington, UN.

UN, 2002. United Nations, European Commission, International Monetary Fund, Organisation for Economic Co-operation and Development and World Bank. Central Product Classification. CPC version 1.1 Draft SA/STAT/SERV.M/77. New York, USA.

UN, 2003. Integrated Environmental and Economic Accounting. United Nations, European Commission, International Monetary Fund, Organisation for Economic Co-operation and Development and World Bank. New York, USA.

UN, 2006a. The International Industry Classification of Economic Activities ISIC Revision 4 Statistical Papers. Agriculture, Forestry and Fishing. New York: United Nations Department of Economic and Social Affairs Statistical Office.

UN, 2006b. Standard International Trade Classification (SITC) Rev 4. No. E.06.XVII. 10. Statistical Papers Series. UN, New York, USA: United Nations, Dept. of Economic and Social Affairs. Statistical Office.

UNECE/FAO, 1987. Conversion Factors (Raw Material/Product) for Forest Products. United Nations Economic Commission for Europe/Food and Agriculture Organization Timber Branch. Geneva, Switzerland.

UNECE/FAO, 2010. Forest Product Conversion Factors for the UNECE. United Nations Economic Commission for Europe/Food and Agriculture Organization Timber Branch. UNECE/FAO, Geneva, Switzerland.

Vinya, R., Syampungani, S., Kasumu, E. C., Monde, C., Kasubika, R., 2012. Preliminary Study on the Drivers of Deforestation and Potential for REDD+ in Zambia. A Consultancy Report Prepared for Forestry Department and FAO under the National UN-REDD+ Programme Ministry of Lands & Natural Resources. Lusaka, Zambia.

ZFAP, 1997. Zambia Forestry Action Plan 1997−2015, Forestry Department. In: MENR (Ed.), Lusaka, Zambia.

CHAPTER 5

Integration of Forestry into the National Economy

Jacob Mwitwa[a], Ambayeba Muimba-Kankolongo[a], Obote Shakacite[a], and Phillimon Ng'andwe[b]

[a]Department of Plant and Environmental Sciences, School of Natural Resources, Copperbelt University, Kitwe, Zambia; [b]Department of Biomaterials Science and Technology, School of Natural Resources, Copperbelt University, Kitwe, Zambia

INTRODUCTION

Integration of the Forestry sector into the national planing process has not being given serious consideration in the past despite its importance in development. In Zambia the focus has been on extractive industries such as mining and more recently on agriculture. Yet, Zambia is an example of a country with abundant and rich forest resources. The forestry sector has many advantages that need to be integrated in the national economy such as:

- *Flexibility*: Scale of operations and technology range from pit-sawing to high technology equipment affording backward and forward linkages, implying important multiplier effects on the whole economy.
- *Remoteness*: Because of their generally remote location, forest industries can create development such as the production of wooden poles for use at household and industry level and provide a wide range of products, including basic necessities, for poor populations such as NWFPs.
- *Import substitution*: Forest products can substitute for expensive imports and can earn valuable foreign exchange when exported. Zambia is a net importer of forest products such as paper and higher-valued wood products. Importation is expected to raise as internal demand is expanding due to increases in consumption as the country's population grows and incomes increase.
- *Renewable resource*: Forests offer a multitude of renewable raw materials for domestic industries and for export.

Forest Policy, Economics, and Markets in Zambia. DOI: http://dx.doi.org/10.1016/B978-0-12-804090-4.00005-7

- *Capital requirements*: The forestry industry is a relatively low capital sector although the labor needs are high in comparison with many other industries. In addition, the investment range is wide, allowing smaller investors to start up businesses such as has been experienced in the sawmilling subsector.
- *Woodfuel*: Fuelwood and charcoal can contribute to economic growth and community fuelwood plantations may be key to increasing production of wood supply for renewable energy.
- *Market demand*: There is increasing demand for timber and other wood products for forestry industries.

However in recent years, national economic development strategies have begun to include the capital and environmental values of forests into national policies, projects and programmes (FAO, 1994; Ng'andwe et al., 2010). Forests are now widely acknowledged as both productive capital stocks and as components of public infrastructure systems (FAO, 1994; UN, 2003) and are entering the central equations of Zambia's macro-economic growth, often with new perspectives (Ng'andwe et al., 2010) of what the forestry sector is and does. In addition, forest systems are part of the resource providing ecosystem services such as water that would otherwise require high capital expenditures and the reduction in population well-being (FAO, 1994). By storing water, regulating water flows, protecting channels and sequestering carbon, forests form a unique ecological and social support structure.

Forest policies in Zambia have evolved out of a narrow sector perspective which has struggled to enter into the mainstream planning process dominated by key economic sectors such as the mining. Because of several emerging issues in the country including climate change (Ratnasingam and Ng'andwe, 2012), forests are currently a major topic of discussion at national and international fora (Banda et al., 2008; FAO, 2010; Chidumayo, 2012; Vinya et al., 2012).

In this context, a significant strain has been placed on the forest policy institution that existed during the last decades. These pressures combined with a wider understanding of the importance and complexity of forest productivity and competing consumption patterns for NWFPs, services, and values, are strongly influencing forestry policy.

FRAMEWORK FOR INTEGRATION OF FORESTRY SECTOR INTO THE NATIONAL ECONOMY

Forestry in the developing world received little or no attention for its contribution to the economic growth during the 1950s (FAO, 1994). According to FAO (1994), development strategies highlighted capital formation and technical progress as the major factors responsible for raising income and economic growth in the developing countries. Generally, forests were viewed as a source of land to be converted to more productive uses rather than for timber and NWFPs. Forests were also viewed as relatively unimportant in the struggle to promote sustained economic development. Several efforts have been made to integrate forestry into the national economy in past years (Ng'andwe et al., 2010). For example, the Zambia National Environmental Action Plan (NEAP) of 1994 provided a framework for making significant policy, legal and institutional changes that were needed to bring environmental considerations into the main stream of decision making. The plan provided an overview of Zambia's environmental problems, existing institutions and legislations, and various strategic options for improving environmental management in the country. Following the NEAP was the Zambia Forestry Action Program 2000–2020 (ZFAP) which provides a range of approaches for sustainable forest management. ZFAP took into consideration the holistic inter-sectoral and interactive approaches to forest management including ecosystem approaches integrating the sustainable use of biological resources and adequate provision and valuation of forest goods and services. It also included components on wood energy and re-forestation activities.

Internationally, efforts have been made to address issues that link forestry to poverty reduction. DFID, FAO and others have prepared agenda of action that have highlighted points of action (Table 5.1) for consideration by international agencies to enhance forest contribution to poverty reduction (FAO-DFID, 2001; DFID et al., 2002). These have reiterated and been complemented by others that are proposed on poverty and environment linkages (Table 5.1).

Some of the above actions have resulted in the formation of the National Forest Programme (NFP) in order to integrate Zambia's forestry in the national planning.

Table 5.1 Agenda for Action on Poverty Environment Linkages	
Agenda for Action from the FAO/DFID Tuscany Forum	**Agenda for Action proposed in the joint DFID/EC/UNDP/ WB document on poverty-environment linkages**
A. **Strengthening Rights, Capabilities and Governance** 1. Support the poor's own decision-making power 2. Strengthen forest rights of the poor and the means to claim them 3. Recognize links between forestry and local governance	A. **Improve Governance** • Integrate poverty-environment issues into national development frameworks • Strengthen decentralization for environmental management • Empower civil society, in particular poor and marginalized groups • Address gender dimensions of poverty-environment issues • Strengthen anti-corruption efforts to protect the environment and the poor • Reduce environment-related conflict • Improve poverty-environment monitoring and assessment.
B. **Reducing Vulnerability** 4. Make safety nets not poverty traps 5. Support tree planting outside forests 6. Cut the regulatory burden on the poor and make regulation affordable	B. **Enhance the Assets of the Poor** • Strengthen resource rights of the poor • Enhance capacity of the poor to manage the environment • Expand access to environmentally-sound and locally appropriate technology • Reduce the environmental vulnerability of the poor.
C. **Capturing Emerging Opportunities** 7. Remove the barriers to market entry 8. Base land use decisions on true value of forests 9. Ensure that markets for environmental services benefit the poor 10. Support associations and financing for local forest businesses	C. **Improve the Quality of Growth** • Integrate poverty-environment issues into economic policy reforms • Increase the use of environmental valuation • Encourage appropriate private sector involvement in pro-poor environmental management • Implement pro-poor environmental fiscal reform
D. **Working in Partnership** 11. Simplify policies and support participatory processes 12. Promote multisectoral learning and action 13. Enhance interagency collaboration 14. Make NGOs and the private sector partners in poverty reduction.	D. **Reform International and Industrial Country Policies** • Reform international and industrial country trade policies • Make foreign direct investment more pro-poor and pro-environment • Enhance the contribution of multilateral environmental agreements to poverty reduction • Encourage sustainable. consumption and production • Enhance the effectiveness of development cooperation and debt relief.
(Source: FAO/DFID, 2001).	(Source: DFID *et al.*, 2002).

Policies and Legislation Relevant to Forestry

National forest resources of Zambia are state controlled through the power vested in the President according to the Forest Act of 1973 Cap. 199, and the draft Forest Act No. 7 of 1999. The policy and legal reforms particularly related to the Forest Act of 1973 including the non-active Forest Act No. 7 of 1999, and other related legislation that are critical in ensuring community access to and benefit accrual from forests under common and protected property regimes. The omission of issues of community governance over forest resources in both the Forest Act of 1973 and the Forest Act No. 7 of 1999 is partly embedded in the utilitarian approach to measuring poverty which is ethically flawed. This approach ignores the condition of life by using externally imposed values.

The Poverty Reduction Strategy Paper (PRSP) and the Third National Development Plan (TNDP) were used as national development strategies from 2002 to 2005. The PRSP constituted a framework for targeting poverty and has been replaced by the National Development Plans (the Fifth National Development Plan, FNDP, which ran from 2006 to 2010; the Revised Sixth National Development Plan, R-SNDP, for 2011−2015; and subsequent National Development Plans which are of 5-year durations). The formulation of the National Development Plans, such as FNDP and R-SNDP, involves the diffusion of the economic, environmental, and social sectors into the Sector Advisory Groups (SAGs) as information sinks for the National Development Plans (NDP) draft process. Forestry institutions as well as the Forestry Department automatically qualified to make submissions and contribute to discussions in the SAGs. In each of the NDP, under Natural Resources, there is a review of the forestry sector's past performance, key reforms, strategies, and objectives. However, these do not provide for the linkage of the forestry sector to poverty reduction but the absence of the articulation of the forestry in some key government strategic sectors is not a reflection of lack of diffusion of forestry, but is a basic government prioritization of focal sectors.

Enhancing Community Access and Benefits from Forests and Woodlands

The right of access for extraction, conveyance and sale of natural property from customary land is also limited by the statutory law and customary norms, as are statutory limitations imposed on forests under state jurisdiction, for example, classified as protected forest areas. In addition, the right of access to forest products as is the benefit arising thereof are critical pre-conditions guaranteeing effective management of benefit accrual from forest resources. Local and national strategies to enhance access and benefit accrual from forest resources hinge on a facilitative legal and institutional framework, access to low cost capital and technology, low local capacity for forest based enterprise development and management, and the absence of linkages and synergies between the forest-based community and market players.

A major consideration is the inclusion in the policy and legislative provisions of devolution of local level governance of forest resources. Because of its apparent physical condition and valuation neglect, the approach does not consider the causes of poverty. The utilitarian approach excludes the enhancement of individual capability to do things differently such as making decisions on forest resource governance, secludes forest-based communities from accessing and benefiting from natural resources through employment that would lead to social insurance and monetary incomes, and also excludes the participation of communities or individuals in the formulation of policies and laws that govern the management of natural resources. Therefore, the absence of devolution in the governance of forest resources to communities has led to social seclusion and inhibited participation of local communities.

Together with devolved governance, clear revenue retention mechanisms are absent in the policies and legislation to guarantee benefits to local communities. The absence of this provision places restrictions on revenue externalization under any partnership arrangement between the state and community, or public—private partnerships (PPP). This would allow adequate growth of social and material capital at local level. Community friendly benefit accrual mechanisms facilitate community access to low cost technology and investment capital through PPPs. These partnerships are important in the transfer of entrepreneurial and business management skills and knowledge to local communities. For these partnerships to function and be beneficial, elaborate policies and regulations that guarantee the safety of private investment from risks must be in place. These risks should be both external or internal such as unfair competition from cheap products and pricing. Restrictive policies, and inadequate and alienating legislation, partly also hinder the roll out of community based projects.

Forest management has often been based on a limited range of economic values omitting other benefits from forest and woodland resources, especially the use of forest products and other ecosystem services by local communities (Ng'andwe et al., 2010).

In 2004, the Zambia Forestry Department partnered with FAO (NFP) in order to improve the forestry-related information exchange, link forest policy and planning with broader national objectives, strategies and programmes, develop partnerships among stakeholders, and build the capacity of the Forestry Department, NGOs, other partner institutions and the private sector in forest policy development and implementation. According to Ng'andwe et al. (2010), the key outputs of the NFP process in Zambia were the generation of information on forest revenue, concession systems, and the contribution of the forestry sector to poverty alleviation and the national economy. This process resulted in the realization that forestry contributes about 5.2–6.3% to GDP per annum (Ng'andwe et al., 2006; CSO, 2008), making the forestry sector the highest contributor to the GDP of Zambia among natural resource based sectors. Without these FAO (NFP) process, estimates of the performance of the forestry sector would have remained at 0.9% of GDP due to the inadequacy of the data, variations in data collection methods, and the lack of information reporting standards and formats.

Support to Forestry Data Handling Program

The nonconsideration of the role of the forestry sector by policy makers in NDPs is as a result of inadequate information on its economics, potential for creating employment, the extent of its contribution to GDP and poverty reduction. This arises out of the fact that the finances provided for the operations of the Forestry Department are inadequate to support the establishment of an institutional framework for forestry statistics acquisition, storage and dissemination. Additionally, this should have been translated into a policy framework that would have allowed the institution galvanize resources to implement and continuously manage the forestry information system. Due to the high cost of the data management program and high human resource turnover, additional support toward capacity building related to management of forestry statistics is critical.

Access to current and reliable statistics on forestry resources such as deforestation, ecosystem services, stocking and allowable species cuts, value addition, trade statistics, available low cost technology and capital related to forestry are indispensable in ensuring effective management and maximal benefits from forestry resources. Issues of access and benefit sharing become relevant when information on the resources to be accessed and benefited from is available. Such statistics, especially with respect to the contribution of the forestry sector to GDP are significant for purposes capacitating policy makers' to make informed decision regarding the allocation of budgetary resources as well as ranking of forestry on the basis of its contribution to the national economy and to rural livelihoods. Creating a forest products data link amongst the

Forestry Department, Ministry of Finance, the Central Statistical Office (CSO), EBZ, and ZRA and other stakeholders such as the timber industry and forestry data gathering organizations is indispensable in capturing the true contribution of the forestry sector to Zambia's national economy. This is likely to enhance investment in forestry as well as policy and legal reforms that are likely to benefit rural communities and other stakeholders.

Handling and disseminating forestry statistics requires appropriate coordination including the validation and duplication unit to manage data acquisition, processing, storage, and dissemination. The data function then interacts with a wide spectrum of stakeholders in the forestry industry, training institutions, civil society, NGOs, and the CSO. CSO together with other stakeholders then prepares data acquisition and handling guidelines, including the understanding of the classification and presentation standards. The standardization of formats and methods are essential for users to apply compatible formats, methods, and reporting periods. Standardized methods include the use of standard units and measures, equipment such as balances, and product definitions. The Forestry Department will then sub-contract or outsource data acquisition to civil society and private firms since it is a regulatory agency and should not carry out data collection to avoid bias. This ensures that data acquisition is supervised by the Forestry Department and is shared in the right standard with other stakeholders.

Dissemination of Forestry Sector Statistics

Forestry Statistics Handbook will have to be published every two years by the Forestry Department and made available to various agencies, stakeholders, such as CSO, and the public. The Handbook could, comprehensively cover a range of forestry sector information including forest areas, carbon stocks, standing tree volumes and allowable cuts, rates of deforestation for each land category and tree species, areas and lands under concession, number of concessions, pit sawyer investments, and tree species being exploited and their harvested volumes and forest type. Additionally, information on NWFPs harvesting and their contribution to household income, numbers of private sector players involved in forestry and their levels of investment, potential areas for investment, formal and informal employment created, land under customary tenure, farm forestry, and the various existing policies and laws are included in the Handbook for use by interested parties and during reviews of National Development Plans.

Diffusion Pathways of Forestry Statistics

Forestry sector statistics—classified into two broad categories, namely statistics related to the activities of the Forestry Department and those involving the downstream privately owned processing industries—reach the various government agencies through the Forestry Department. Some of the activities of the Forestry Department involve human resource development and employment, forestry related revenue generated from licenses and fees, and externally funded project related statistics. Activities from the private sector should also diffuse into the planning process through several other statutory bodies such as the Export Board of Zambia, CSO, and the Zambia Revenue Authority (ZRA). Information on licenses and fees covers human resources, number of concessions and pit-sawyer licenses, fees collected for the extraction of wood and NWFPs, as well as statistics related to the extent of the state plantations for *Pine* and *Eucalyptus* species.

Generally, the Forestry Department prepares the annual report that is submitted to the Permanent Secretary of the relevant Ministry dealing with Environment and Natural Resources. The report covers all activities of the Department including revenue and expenditure that also shows the forecast of the next financial year's estimated revenue and expenditure. Based on the information available, the annual budget is prepared as part of the Ministry's budget and submitted to the Ministry of Finance and National Planning. This budget is often an aggregate of the budget estimates prepared by the various divisions of the Department from district to the Province. Since the district and provincial forestry offices are part of the local government system, District Development Coordinating Committee (DDCC) and District Natural Resources Committee, the annual budget submitted to the Forestry Department main office in Lusaka is also reflected in the local government structures.

Cabinet circular No. 1 of 1995 established the institutional framework for planning involving the creation of Development Coordinating Committees (DCC) at National, Provincial and District levels. In the District, the decentralized development planning process is the mandate of officers of line departments, Civic officers, donors, NGO who are coordinated by the Town Clerk to form the DDCC. Functions of the DDCC are as follows:

- Co-ordinate District planning and project implementation across Council and line Departments;
- Provide the link between NGOs and District Councils;

- Consolidate District plans for approval by the Councils;
- Receive and assess project proposals from communities and Development Agencies in the District;
- Co-ordinate the preparation of and implementation of annual capital programmes;
- Co-ordinate and Monitor District Project implementation;
- Evaluate completed projects and review District development plans;
- Monitor and Co-ordinate Sub-district community participatory planning activities;
- Prepare consolidated reports on District development for the council with copies to the PDCC and other relevant National Institutions.

Therefore, district budgets and annual plans are submitted to and incorporated as part of the strategic development plans of the specific districts that are then submitted to the Provincial Development Coordinating Committee (PDCC). After the budget is made public by the Ministry of Finance and National Planning, the Forestry Department is invited by a Parliamentary Committee to explain and validate related specific estimates. The approval of the annual budget by Parliament is a direct endorsement of the Forestry Department's annual plan of activities. In general, forestry sector statistics have been collected by EBZ, CSO and ZRA and often include reports on volumes of exports, the contribution of the forestry sector (as part of the agriculture sector) to GDP, and revenues collected from the private sector in form of taxes. Therefore, such linkages and partnerships between these institutions should be streamlined and strengthened to allow for a constant data flow between them.

Forests as a Source of Income and Revenue

Income and revenue from forestry derive from direct payment of forest fees, royalties, charges, and issuing of forest licenses for the cutting, conveyance, and utilization of forest products. Generally however, only very low revenue accrue from the Forestry Department related activities mainly because of a number of factors including ineffective revenue collection system, low price of the products, poor funding of the Forestry Department, lack of proper control and monitoring of resource exploitation and poor law enforcement, illegal logging, and lack of capacity to supervise and monitor forest operations (FAO, 2003a, 2003b). Consequently, the annual revenue collected by the Forestry Department represents only about 12% of the potential (ZFAP, 1997). The collected revenue is channeled to the central treasury and only a 45% share is retained annually as appropriation in aid to support forest operations. The rest is shared between the Ministry of Finance and National Planning (30%) and the Ministry in which FD operates (25%) (Chileshe, 2001).

Prices of Forest Products
The Statutory Instrument No. 121 of 2003 stipulates the fees and prices for all forest products, but these are regularly reviewed by the Forestry Department. The fees to be paid for a product are expressed in fee units and to get the actual price in Zambian Kwacha (ZMW), the number of fee units is multiplied by 180, a multiplier factor used by all government institutions for charges of the goods and services.

However, the determination of prices for forest products for revenue collection purposes are not based on the analysis of estimated cost and revenue structures of the different products. This has led to regular recommendations for price increases that have allowed marginal operators in the sector to survive while more efficient operators make more profits. In addition, the setting of fees and prices has often resulted in a weak revenue collection system making recommendations for any increase difficult and always resisted by some timber processing industries and associations.

Handling of Forest Revenue
All forest officers have authority to inspect consignments of forest products, search, verify quantities and identity of holder of forest permits, validity of licenses, source of forest products and destination, and list and type of products before the products enter the market or the sawmill site. The collected fees are registered in the various documents such as license books, revenue receipt books, revenue forms, and report forms before depositing in the bank. Record keeping including planning and audit checks are carried out as per laid down by government guidelines and procedures. Forest officers issue forest licenses and permits for the production, conveyance, and sale of the forest produce.

Monitoring of License Compliance
Licensing of forest products is used as a means for monitoring and controlling the exploitation of forest products and in many cases, the control is remotely done (Whiteman and Brown, 1999, Whiteman, 2004). In view of this, various products from forests in Zambia are unsustainably harvested. Illegal harvesting of the forest products like

charcoal, firewood, and timber is common and often in violation of the forest laws and subsequent regulations. It is estimated that illegally harvested wood for various purposes, for example, far exceeds the official annual wood production of over 13 million m^3 per annum.

The underlying causes of illegal operations are many including a flawed policy and legal framework (Mwitwa et al., 2012) inadequate law enforcement by the Forestry Department, insufficient data and information about forest resources, and the rampant corrupt practices for law enforcement officers. Furthermore, illegal trade in forest products is also associated with other illegalities such as tax evasion and corrupt practices, thus leading to losses of revenue for the Forestry Department. The implications are long term negative economic and environmental impacts as a result of unsustainable harvesting, deforestation and loss of tax revenue. Illegal harvesting of forest products also has jeopardized the livelihood of rural communities particularly those engaged in small-scale forestry activities by exposing them to unfair competition and depleting resources on which they depend. Therefore, any attempts to promote sustainable forest management through investment will remain ineffective unless these problems are addressed (FAO, 2005).

Joint Forest Management (JFM)
The legal and institutional framework in the country ensures that the state control over natural resources remains central to forest resource management. Even though existing legislation anchors centralized management, the Forest Policy of 1998 encourages a participatory approach in the management of forests by promoting partnerships between state agencies, local communities and individuals. Paradoxically, the Forest Act of 1973 Cap 199 still maintains state control and ownership. Despite the passing of Statutory Instrument No. 52 of 1999, significant weaknesses related to resource access rights, benefit and tenure rights within the forest resource governance infrastructure of Zambia.

In an attempt to reform policy and legislation, the forestry regulatory framework has attempted to include a clear definition of who the beneficiary is. These stakeholders include forest-based communities that are heavily dependent on the forest resource, communities living outside the forest but drawing heavily on the proximal forest resource, individual owners of private forests such as farmers, and processors and traders in forest products including middlemen. Community participation in the management and governance of private forest property regimes, which are not provided for adequately under existing legislation, are guaranteed access, resolution of conflicts over use, and can access private sector investment. Devolution of forestry resource management to local communities and allowing the creation of robust local institutional frameworks that are guided by local by-laws is expected to provide a framework for sustainable forestry resources management. The state would still maintain a regulatory responsibility to provide checks and balances that ensures that national forestry resources are sustainably managed.

Nevertheless, the amount the Forestry Department retains as 45% from the revenue collected from forest activities is inadequate to meet all subsequent costs associated with sustainable management and protection of forests. Since 1995 however, the Forestry Department undertook a pilot participatory forest management system in Eastern, Luapula, Copperbelt and Southern Provinces as an alternative to the centralized forest management practice. The rationale for promoting this approach was based on the assumption that effective management of forest resources would be attained when local communities, who are the main users of the forest resources, have a share and legal rights to make decisions and obtain benefits from the use of such resources. This has partly been driven by the failure of centralized management and global changes in the management of natural resources. In 2005, the Forestry Department in collaboration with local communities in selected pilot areas prepared a Joint Forest Management (JFM) plans through the PFAP to promote sustainable forest development. However, the implementation of the plans was delayed due to legal implications even though the Statutory Instrument No. 47 of 2006 empowered the local communities to jointly manage forests in areas declared as JFM areas and to share revenue. The source of revenue from JFM areas includes major forest products such as timber building poles, charcoal, and firewood. However, gaps in legislative provisions detailing benefit sharing mechanisms comparable to what exists in CBNRM under wildlife management ensured that JFM could not take off and be a successful forest management system.

Cost and Benefit Sharing of Forest Revenue
The benefit sharing of the accrued revenue from forest resources is still in its state of infancy considering that the Forestry Department has not prepared the legal framework, even though it is under active consideration, on how to share these benefits among the key stakeholders. It is proposed for the communities to receive 70% and 30% to be for the government.

However, it is unclear from this model how the allocated proportions will be shared among the various local stakeholders, constituting a likely source of conflict. But, most of the communities in the JFM areas have indicated that the shared revenue will go toward community development projects such as schools, health centers, roads, sanitation and water supply, as well as to meet other forest resource management costs. This is also similar to what happens in Community Resource Boards in Game Management Areas when revenue from ZAWA is received. Investing such small amounts of money in social services, which is the role of the government, has little impact on the community and the effectiveness of resource management. This is more apparent in years when there is no revenue accruing from resource management. The most effective model is to create a seed fund under which the investments made continuously generates revenue for the community. The profits from such an investment are then allocated to the support of developing social services. Forest resources in Zambia remain low value resources compared to wildlife. Expecting to raise large sums of money from such low value resources is a fallacy that may affect participatory management systems. Only in areas where the forest to be managed is comparative larger and the stocking of economically valuable species is higher. The species include timber species, medicinal plants, plants used for food such as wild vegetables, and thatch. Higher revenues can also be enhanced by other ecosystem services and enterprises such as tourism and cultural villages. These should be built around JFM in partnership with the private sector.

The private sector has the potential to increase the contribution of forestry to the economy, both national and local, especially when JFM is considered. Even if revenues accrue to the local stakeholders, the low local education and capacity cannot be expected to enhance the multiplication of the revenues received. Most of it is likely to go into consumption and not investment for profit.

Contractual Arrangements in Forest Licensing System
As highlighted above, several challenges characterize revenue collection in the forestry sector. As a result, some contractual arrangements have been put in place and these include the followings:

1. Sawmilling and pit-sawing are awarded only to companies and to individuals/groups having sawmilling or pit-sawing capacity. This

approach reduces competition and prevents the development of independent logging operations. However, the consideration is based on preventing corrupt practices and collusion by individuals who would be awarded contracts and either fail to utilize them or re-sell the same to others who have proven capacity.

2. Sawmilling and pit-sawing are awarded through negotiations rather than auctioning while concessions are awarded at the discretion of the Forestry Department, often with political interference, persuasion, and kick-backs.

3. Producers are required to supply only the marketable products (e.g., saw logs) leaving other tree parts in the production and processing processes to rot in the forest resulting in considerable loss and underutilization of the resources. Additionally, this increases the biomass on the forest floor which results in increased gas emissions during the fire season.

4. Forest concessions are awarded in protected and customary forests after prior informed consent for the local community, forest inventory, and other requirements are put in place.

5. Advance royalty fees are paid for high value timber species.

The current system of forest concessions and pit-sawing licenses has several drawbacks that include:

1. Most forest concessions are awarded without properly executed forest inventories resulting in inadequate and poor planning of concession activities.

2. Prior informed consent is by "informing the community and its leadership about the intended forest concession" and rarely a discussion to obtain the consent of the community. Communities in whose areas, particularly customary forests, the concession is awarded are not informed of both the environmental and economic costs of the concession. These include loss of rights of access to the concession area. Long-term environment impacts related to forest degradation include loss of important medicinal plants and plants used for food as well as mushrooms.

3. The advance royalty fee payment system currently in use does not reflect the real value of the forest products. Even though concession licenses are based on the production of high value timber species, there are no incentives for collecting forest revenue from low value timber species.

4. Concessionaires do not participate in any reforestation program and do not contribute to reforestation funds, making the terms and conditions of concession licenses inadequate. Local communities do not benefit from this activity, and the people employed are usually from outside the local community. Migrant labor impacts on the local community economically and socially. Concessions are not accompanied by any activity that empowers the local community to effectively deal with these external influences.

5. Forest officers are in most cases compromised by the failure of the Forestry Department to provide funds for concession management. The concessionaire normally provides transportation, lunch and other related costs to the forest officer for the conduct of the forest inventory, consultations with the local community and local authority and monitoring of the concession. The allowances "unofficially" paid by concessionaires are usually better than those paid by the government—this is paid on the spot without the hassle of completing a multitude of official documentation.

6. Monitoring of harvests and conveyance: Timber marking hammers are rare in some of the concessions, therefore paint or charcoal is used to mark the timber. This has contributed to illegal harvests as the timber cannot be traced to any concession. The Forestry Department is poorly financed and has been so poorly managed since the 1990s that it is one of the government departments that has no transportation to monitor timber extraction and transportation. Concessions in Northwestern Province for example, are monitored only when a senior official is visiting or when the concessionaire desires that a monitoring visit is important. Transportation is then organized by the senior official or the concessionaire.

In order to address the many challenges facing the management, revenue collection and monitoring of activities taking place in the forest, there is need for improved capacity in the Forestry Department, the development of clear strategies and procedures for the allocation of concessions, and the creation of assurance of rights and obligations for the concession holders and stakeholders such as:

1. Development of a clear strategy for identifying timber production areas which would be managed as forest concessions in line with forest certification standards on a sustainable yield basis. Such areas may be blocks of clearly defined forest areas or groups of small blocks in the forest areas.

2. Allocation of concessions by sealed tender or competitive bidding based on per hectare bonus bid payable annually on the total area of the concession to promote competition in the whole national timber concession system.

3. Promotion of annual concession systems which reflect area-based concession fees at rates that have the potential to generate a significant proportion of forest revenues and provide incentives for forest management. Annual concession fees would be a major source of revenue and should supplement or partly replace the most difficult-to-collect volume based stumpage price and conveyance fees.

4. Introduction of a system of *difficult to alter* labeling system particularly for commercial charcoal production that ensure a more effective system of controlling product production and revenue collection. A label will be designed such that it includes district name, serial number and explicit instructions which should always be placed on the bag of charcoal and remains there to trace the channel of the production site to the market and the end user. This is feasible in commercial charcoal production and not in the small-scale and illegal charcoal production system.

5. Introduction of other types of licenses such as charcoal production license, forest concession license, sawmill license, pit-sawing license, conveyance license and firewood license to avoid the confusion from the previous type of casual license covering all types of products. When a license is issued, the licensee is provided a complete set of all relevant rules, regulations and other instructions relevant to the operation of the license. Guidance is also given to field staff regarding the interpretation of forest law and regulations.

6. Initiation of a minimum volume-based stumpage price sufficiently high enough to reflect the administrative costs of supervision, monitoring, forest concession renewal and forest management, scaling and collection of revenue, environmental and other nonmarket values, and the opportunity cost values that are not included for harvesting the timber. These minimum volume-based fees can prevent below cost or below opportunity cost harvesting, hence improving the overall efficiency of forestry revenue collection.

7. Establishment of a fund to finance forest management, supervision and monitoring of forest concession licenses. This fund, about 30% of the forest revenue generated from sawmilling and pit sawing concessions, could be allocated for forest management activities of concession areas and be used to support the supervision and monitoring of harvesting of timber.

8. Revision and alignment of the current licensing conditions with inter-
   nally accepted forest certification principles, criteria and procedures.
   In general, the contract to be offered should provide certainty of rights
   and obligations for the concession holder. The contract performance
   conditions are specified with clear steps, and provision of incentives
   for compliance with the management obligations by the concessionaire. These conditions must be emphasized on ground performance
   and ensures reliable and sustained monitoring and verification.

LINKING THE FORESTRY SECTOR TO POVERTY REDUCTION

The UN MDGs as interpreted in the National Development Plans
(NDP) commit the country to halving the proportion of its citizens living
in absolute poverty and the proportion of people suffering from hunger.
In Zambia, poverty is principally measured from the utilitarian basis as
deprivation of income which in principle focuses on the subsistence of a
family unit of six or simply a household. A food basket necessary to feed
a household of six is determined and all households with income which
cannot purchase food equivalent to the estimated food basket are considered to live in poverty. Like many other countries in Sub-Sahara Africa
which are highly dependent on subsistence agriculture, Zambia faces
numerous challenges of surviving in an increasingly "globalized" economy, and all the factors of international trade which are not supportive
of rapid economic growth. Particularly, Zambia's economy remains
highly dependent on copper resources, which has the tendency of price
fluctuations. In attempting to arrest and reverse the economic decline,
the Zambian government implemented a sweeping programme of liberalization and deregulation during the 1990s aimed, inter-alia at eliminating
most major market distortions and allowing the private sector to develop
the economy. The most affected location was the Copperbelt Province as
this remains the most industrialized region in Zambia.

However, commitment to reform weakened in the mid to late 1990s
when macro-economic stabilization led to an initial contraction of the
overall economy. The Poverty Reduction Strategy (GRZ, 2002) and the
Transitional National Development Plan (TNDP) sought to achieve
poverty reduction through promoting economic growth and supporting
infrastructure development, improved governance, improved access and
quality in the provision of social and public services, and mainstreaming
of cross-cutting issues such as HIV and AIDS, gender equity and the

environment. Despite the economy being copper driven, about 50% of Zambia's population live in rural areas and less than 10% of them are employed in the formal sector. Rural communities are principally peasant in nature, with subsistence agriculture as the main form of sustenance combined with various levels of dependence on forests, freshwater or wildlife.

The potential role of forestry in alleviating poverty has been accepted and is receiving considerable attention worldwide and locally. In Africa, Uganda was perhaps the first country to have developed a poverty reduction strategy in relation to forestry under the National Forest Programme (Margaret et al., 2006). Uganda went through the NFP process during which considerable efforts were put in place to influence the Poverty Eradication Action Plan (PEAP) and align clear strategies for the Forestry sector with its pillars in the overall country development framework. The process resulted in the development of the PEAF report that outlined details for the sector's contributions to the national economy, the problems it faced, its potential for poverty alleviation, and several external influencing factors such as land ownership, energy consumption, and decentralization and urbanization. According to Margaret et al. (2006), the current PEAP in Uganda shows that forests provide an economic value of USD360 million (6% GDP) per annum of which only USD12 million is captured in the official statistics. After 3 years of persistent lobbying by stakeholders, PEAP in Uganda now regards forestry not only as a sector, but as an "urgent short-term priority" that is considered for increased funding from government and other stakeholders. The rationale is that a large proportion of the population derives a living, or part of their livelihood, from forest resources. Therefore, enhancing the role of forest resources in poverty reduction programmes is both a sound strategic issue as well as economically feasible compared to programmes such as employment creation.

Role of Non-Wood Forest Products in Poverty Reduction

Forests and trees constitute one of the major sources of rural income and employment. They provide for subsistence needs and reduce vulnerability of the population. A wide range of commercial products also originate from forests such as round-wood for manufacturing some of which are used in copper refineries, pulpwood and fuelwood. These generate revenue for the government and local communities through employment in firms that process these. One of the key components of the contribution of forests to the livelihoods is its carbon sequestration and catchment properties that have a significant contribution to the maintenance of hydrological cycles. Rural communities produce and market a number of wood and NWFPs from forests. Many of these products are labor intensive and are sold from homes, at community markets and along the roadside while some are transported and sold in urban areas. These economic activities can be linked to poverty reduction in both urban and rural areas in forest dependant households (Figure 5.1).

In the case of NWFPs there is a fascinating linkage to poverty alleviation in the way households utilize these products as a source of livelihood. Ng'andwe et al., (2006) compared the NWFP value added with the poverty levels in each Province and found that in most Provinces including Northern, Southern, Western and Central with higher incidence of poverty the value added was lower. This finding was linked to poor households who gather NWFPs mainly for consumption rather than trade and due to possible depletion of NWFPs stocks from forests. The finding was consistent even when analyzed at district level (Figure 5.2).

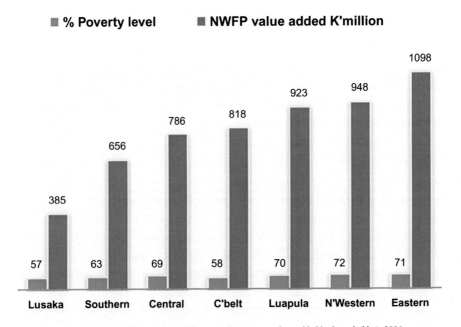

Figure 5.1 Linking poverty levels to non-wood forest products gross value -added by households in 2006.

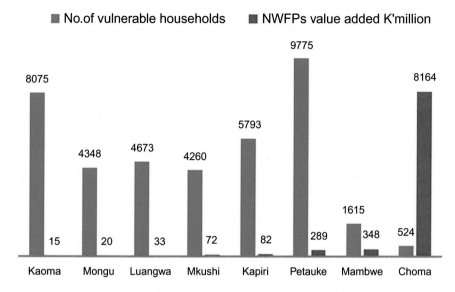

Figure 5.2 Linking vulnerable households in the various districts of Zambia to non-wood forest products value added.

Figure 5.2 also illustrates that the NWFPs value added by households is considerably lower in several districts such as Kaoma, Kapiri Mposhi, Mkushi, Mongu, and Petauke. These were also the most vulnerable households identified by ZAVC (2003) where the population is in constant need of food support from government. Apparently, either most of the collected NWFPs are also used for home consumption and only a little share is carried to market places for sale or they have been considerably depleted due to continuous harvesting as a result of regular food shortages. On the other hand, value addition is higher in households which are less vulnerable such as in Choma. Figure 5.2 illustrates how rural households, who are also among the poorest in the country, rely on NWFPs as a source of food, medicines and income to sustain their lives. NWFPs are mostly used for domestic consumption by households in the informal sector, a view which corroborates well with that of several other authors (Chishimba, 1996; Mogaka et al., 2001; Guveya, 2006; Puurstjärvi et al., 2005; Ng'andwe et al., 2006).

A cross-tabulation reveals more of such linkages in detail (Table 5.2). For example, areas with low poverty levels and low value added to NWFPs show weak income generating activities from NWFPs compared to areas with low poverty but with high value added where income generating dominates (Table 5.2).

Table 5.2 Linkage Between Non-Wood Forest Products and Poverty Alleviation

Factor	Type of Linkage	
	Low Value Added	High Value Added
		market places.
	∞	∞
	Wealthier groups not in need of food aid in rural areas (e.g., teachers, nurses, traders)	*Wealthier groups not in need of food aid in urban areas (e.g., middle men traders)*
LOW POVERTY	• Alternatives exist for consumption and livelihood • Other income generating activities exist • Low prices of NWFPs • Trade in NWFPs highly seasonal	• Income generation activities dominate • Households do not rely mostly on NWFPs for survival • Formal employment exists • Supply of NWFPs from forests exist all year round • Households mostly traders than producers
	∞	∞
	Weak income generation from NWFPs— many alternative products exist	*Strong income generation than consumption*
HIGH VULNERABILITY	• Constant food deficit • NWFPs ensure food intake • High prices of other food commodities • Livelihood unsecured • Households help each other for livelihood • NWFPs depleted • Markets not always available • Households in persistent need of food aid and with no capacity to sustain subsistence	• Constant food deficit • Only NWFPs ensure food supply and counter-seasonal sources of income • NWFPs available all year around • Constant reliance on NWFPs for livelihood • NWFPs mainly processed for value addition for market high prices • NWFPs only option for income generation
	∞	∞
	Very poor groups mostly in remote rural areas (e.g., weakened very old, permanently ill and destitute young aid-orphan households)	*Low-income groups or marginalized communities in urban areas (e.g., young and very productive headed-households, female-headed households)*
LOW VULNERABILITY	• Food intake ensured all year around • NWFPs traded only to fill particular income gaps or needs • Low income and increased poverty • Other better livelihood options available • Markets for NWFPs available	• Other options available for livelihood • Part of livelihood derived from NWFPs • NWFPs safety net only during shocks as famine from natural disasters • Growing demand for NWFPs from markets • Good prices of NWFPs • Transport systems available to reach

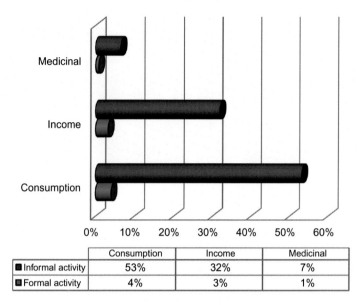

	Consumption	Income	Medicinal
■ Informal activity	53%	32%	7%
■ Formal activity	4%	3%	1%

Figure 5.3 Frequency of non-wood forest products utilization either as formal or informal activity among the households in the different areas of Zambia.

In 53% of households in the informal cluster utilize NWFPs for consumption compared to 4% in the formal cluster. Correspondingly, 32% use NWFPs for income generation, 7% for medicines while in the informal cluster the proportions are 3% and 1% respectively (Figure 5.3).

Changes in the economic environment, with the adoption of a liberalization program in the last decades, have adversely affected the income earning opportunities of most households especially those in the remote parts of the country. These communities have resorted to using forest resources as their last means of survival when they face food shortages and for income generation. Overall, poverty levels are high in most remote rural areas and this is exacerbated by a considerable lack of access to capital assets such as credit facilities and saving arrangements. Those in urban centers, however, have access to several employment and trading opportunities available to meet their food requirements and other socioeconomic needs. However, because most of their purchasing power has considerably declined during the past few years more especially as a result of loss of employment, many households in urban

Figure 5.4 Selling of non-wood forest products has become an increasing business among many households to earn income: (a) Thatch grass, and (b) caterpillar in urban market.

centers have taken up informal employment and petty-trade as their main sources of livelihoods. Petty-trade is current particularly where opportunities for cross-border trade and tourism are high. Common items traded include foodstuffs, handicrafts, livestock, and consumer goods. Furthermore, due to the prevailing drought in some areas of the country like in Southern Province, poor households rely only on wild resources from forests for food and for income generation (Figure 5.4).

Role of Forestry in Employment Creation

Great concern over poverty levels in the developing world has been expressed by leaders of the developed world and many of the UN affiliated organizations. The development of strategies to meet MDGs is a case in point aimed at eliminating or reducing poverty in a coordinated manner throughout the world (Kabananukye et al., 2004; Banda et al., 2008). Kabananukye et al. (2004) explained that countries that have succeeded in significantly reducing poverty have also high rates of economic growth. They further noted that high economic growth is not necessarily a sufficient condition for poverty reduction but the pattern and sources of growth as well as the manner in which its benefits including employment opportunities are distributed, are equally important in achieving poverty reduction.

The gathering of NWFPs is the largest employer accounting for 83% of the total informal employment seconded by charcoal and fuel-wood collection. On the other hand, the number of individuals setting up new small-scale commercial forest based enterprises and sawmills engaged in both hard and soft woods processing is on the increase. On the Copperbelt for instance, the total number of small-scale sawmills is estimated at 500 and about 1,000 country wise. These small-scale producers employ on average about 20 to 22 persons per sawmill. The size of potential employment including its multiplier effects is quite high. In forest plantations where timber harvesting activities are undertaken, women provide a range of services including selling of foods and beverages to loggers and transporters. These activities have become a common source of income for most women living around nearly all the forest plantations in the country. This clearly indicates that the indirect contribution of the forestry sector to social development of Zambia through employment creation is significant.

Gender in Forestry and Its Relation to Poverty Alleviation

In Zambia, all members of households (e.g., man, wife, children, and other dependents) are generally involved in the collection of forest products. Nevertheless, some products such as thatch grass, *Raphia farini-feri* (Ifibale), Rattan and bamboos are often collected by men but women and children are only involved in the transportation. Females, on the other hand, are actively involved in the collection of NWFPs for subsistence reasons and not for profit. They collect forest products such as mushrooms and caterpillars which are then carried out either to the village or to market places. However, in the downstream processing of forest products, the males constitute the major human capital involved in generation value added.

In 2005, Nguvulu (2005) had also shown that trading of especially wild fruits in the country was mainly a task performed by females. Harvesting of other products like lusala (*Dioscorea hirtifoli)* tubers seems to be tedious more especially that the crop is mixed with several other forest trees in addition to the tubers being deeply buried and anchored into the soil in forests (Nguvulu, 2005). Because of this, both women and men partake this activity which involves looking for the crop in the forest and when this is found, pulling and digging out the tubers in very harden soils. Similarly, Mulombwa (1998) had previously reported men, women and children to be involved in harvesting most of forest vegetal products in the country.

Generally, women play a key role in Zambian society at different levels including as members of the labor force, producers of both marketed and unmarketed goods and services at households level and as an important and effective manager for savings to accumulate the

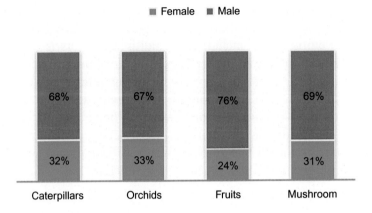

Figure 5.5 A precentage comparison of males and females involved in gathering non-wood forest products.

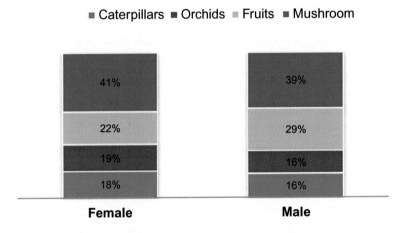

Figure 5.6 Proportions of quantities of non-wood forest products collected by males and females in Zambia, 2006.

household's capital (Mulombwa, 1998; Nguvulu, 2005; Ng'andwe et al., 2006). Women are the tillers of land, the gatherers of food, the collectors of fuelwood, fodder and water, and caretakers of their families. While women perform both productive and reproductive tasks, their role is often disregarded even if their roles have resulted in negative impacts upon their time, income, nutrition, and health, and their contribution to developmental activities and to community managing roles. Additionally, this also impacts upon their access to, and control and management of natural resources. They participate fully in all the activities of the forestry sector right through the product value chain for NWFPs. Figure 5.5 and 5.6 compares the economic activities of gathering NWFPs by women and men. Men dominate in the gathering of NWFPs such as orchids, wild fruits, caterpillars and mushroom for income generating.

More importantly, the forestry sector constitutes a great enterprise acting as one of the major sources of informal employment for a great proportion of households in the country. It contributes significantly to poverty alleviation particularly in rural areas. It is estimated that more than 60% of the economically active population earn their livelihoods from forest related activities in Zambia. With conducive government policies, other forest dependent industries could be created and consequently increase employment opportunities for the population. Furthermore, policies aimed at supporting rural entrepreneurship would have a multiplier effect on self-employment in the urban areas for the unemployed youth and women who eventually sell or add value to forest products from rural areas.

Community Empowerment in Forestry Management

Local management systems that are strengthened by devolved rights and governance frameworks are effective scenarios for effective management of common property regimes. Restrictions on access and governance of state controlled forest resources are partially replicated to common property resources.

Hence though the empowerment of local communities in the management of forest resources is necessary, it can only be effective when an institutional dialogue framework for dialogue and collaboration is in place. This will provide additional sub-sector frameworks, within the forestry sector, with specific discourses on issues related to the management of natural resources. Such frameworks ensure that community interests are primary especially as related to community rights (i.e., tenure, access, alienation) and benefit accruals.

Security of Tenure in Forestry

Local communities are subject to various shocks related to forest resources availability which is due to several factors including climate change. For example, some beekeepers express worry that sometimes the honey season is very poor due probably to the heavy rainy season or a combination of factors. A reduction in liquid honey production from 250 liters to less than 100 liters per capita, a 60% drop, due to climatic and other factors is very significant especially when this constitutes a major source of income for the household. Currently forests on customary lands are open pool resources and are "owned" by the community without the security of tenure which is a title. This factor is compounded by the fact that forests can be used *in situ* for personal consumption but cannot be traded. Extraction of forest products, including timber, must be facilitated by a license from the Forestry Department even if such products originate from a customary area. This hinders investments in community forest projects and is one of the factors that might affect community forest carbon projects unless significant legal reforms are carried out.

Opportunities for Community Benefits from the Forestry Sector

Weaknesses in the position of the poor preventing the exercise of other available options that a resource endowment could provide are key examples in the lack of extended utilization of several opportunities existing in rural areas. Such opportunities could be like the formation of commodity organizations that could control local and regional markets for Nwfps. Part of the cause of the lack of such initiatives is often the inability of the government to create and provide necessary incentives for the formation of commodity groups and/or public—private partnerships in the management of forest products. The formation of commodity groups or cooperatives focusing on forest products empowers a local community grouping to enhance product competitiveness through the establishment of marketing strategies, and access to low cost capital and technology to enhance forest product qualities and marketing. Local communities still compete at low scales that do not encourage the creation of innovations in product quality and marketing rendering commercially viable forest products as poverty traps instead of poverty alleviation opportunities.

Public—private partnerships constitute key socioeconomic instruments in bringing a strong business planning approach to forest management which can lead to the development of effective strategies for using limited forest resources to achieve maximum impacts on livelihoods (GRZ, 2009). In order to significantly contribute to poverty alleviation particularly in rural areas, strong emphasis should be on building public—private partnerships and improving the policy framework for such partnerships. Public—private partnerships emphasize business planning and sustainable financing for state controlled forest resources, common property regime and the socioecological system.

CONCLUSION

The contribution of the forestry sector to poverty reduction and the national economy is significant even though there has not been a major investment in the sector compared to other sectors such as the mines and agriculture. Due to its low apparent contribution at the macro-economic level, the forestry sector has not been able to establish itself as an important element in the development processes of the country as compared to other sectors of the economy like mining, manufacturing, tourism or agriculture. The predominance of such as wood energy and timber processing and lack of integration suggest that the forestry sector still suffers from lack of investment due to political, economic, policy, and structural problems. Still the contribution to poverty reduction and the national economy is high due to other benefits derived from forest resources apart from monetary aspects. Although environmental and other related services are being recognized today, wood and wood products in the manufacturing sub-sector will remain a major source of formal employment in the future—a compelling reason for government to invest in the sector for the purpose of realizing its full potential and also by moving the informal processing up the value chain. In a nutshell, despite the lack of meaningful investment in the sector the contribution of the forestry sector goes beyond the narrow domain of economic benefits in monetary values, it requires a re-look on national income estimates, creating forestry sector links to poverty alleviation, establishing a payment for ecosystem services framework, and developing alternatives energy sources.

The framework for integrating the forestry sector into the national planning process include constant reforms in the policies and legal frameworks. To achieve full integration in the planning process, stakeholders and govenrment planners are expected to develop pro-poor policies based on the judicious exploitation of the indigenous forest resources in Zambia. This may include the strengthenening of the institutional and policy framework governing the practice of JFM. Among the reforms to the policy related to JFM as provided for under Statutory Instrument No. 52 of 1999 are (i) benefit sharing mechanisms should follow a "Bottom–Up" principle as opposed to the current "Top–Down" principle. This will be more desirable as an incentive and a clear practical demonstration of local community ownership and governance over their resources. The integration of JFM with on-going development programmes starting with linking JFM to national development processes such as the PRSP, the NDPs and Biodiversity Strategy and Action Plan.

Additionally, the base benefit sharing has to be integrated, on the principle of differential benefits, whereby communities living with the resource and bearing a higher cost should receive higher benefits than those that do not bear this cost.

Approaches should have a clear elaboration of community rights, responsibilities, ownership and governance of the forest resources as well as the land on which the resources reside which may include aspects that could enhance the linkages of the sector to poverty alleviation such as:

1. public–private partnerships that will build enterpreneurial and resource management capacities at local level resulting in increased capitalization of local level commodity production and marketing;
2. stakeholders and government collaborative frameworks that promote the establishment of cooperatives or the development of rural entrepreneurs in order to enhance employment creation and sustainable income generation for rural and urban areas. Supporting rural entrepreneurship will have a multiplier effect on self-employment in the urban areas for the unemployed youth and women who eventually sell or add value to some of the wood and NWFPs;
3. promotion of informal sector employment that is based on NWFPs through tax incentives and the provision of extension services for sustainable production of NWFPs, domestication and market identification;
4. deliberate rural women empowerment programmes that are planned for and based on the sustainable exploitation of wood and NWFPs on commercial basis through JFM initiatives.

REFERENCES

Banda, M.K., Ng'andwe, P., Shakacite, O., Mwitwa, J., Tembo, J.C., 2008. Markets for wood and non wood forest products in Zambia. Final report submitted to FAO-NFP.

Chidumayo, E.N., 2012. Development of Reference Emission Levels for Zambia. Report prepared for the UN Food and Agriculture Organisation (FAO) and UN Reducing Emmissions from Deforestation and forest Degradation (UN REDD). Makeni Savanna Research Project, Lusaka, Zambia.

Chileshe, A. 2001. Forestry Outlook Studies in Africa (FOSA) — Ministry of Natural Resources and Tourism — Zambia. Available online: <http://www.fao.org/forestry/FON/FONS/outlook/Africa/AFRhom-e>.

Chishimba, W.K., 1996. In-Depth Study of Consumption and Trade in Selected Edible Vegetal Non-Wood Forest Products in Central, Copperbelt and Luapula Provinces. Provincial Forestry Action Programme. GRZ, Ministry of Environment and Natural Resources, Forestry Department, Department for International Development Co-operation, Ministry for Foreign Affairs of Finland. Ndola, Zambia.

CSO, 2008. Central Statistical Office Revised Estimates Real. Ministry of Finance and National Planning, Government Printers, Lusaka, Zambia.

DFID, EC, UNDP, WB, 2002. Linking Poverty Reduction and Environmental Management— Policy Challenges and Opportunities, July 2002.

FAO-DFID, 2001. How Forests Can Reduce Poverty. Policy Brief, November 2001.

FAO, 1994. Food and Agriculture Organization of the United Nations, The State of Food and Agriculture 1994, FAO Agriculture Series, No. 27, ISSN 0081-4539.

FAO, 2003a. Forest Outlook Study for Africa: Sub-regional Report-Southern Africa. Development Bank, European Commission and FAO, Rome, Italy.

FAO, 2003b. An Illustrated Guide to the State of Health of Trees: Recognition and Interpretation of Symptoms and Damage. Food and Agriculture Organisation of the United Nations, Rome, Italy.

FAO, 2005. Global Forest Resources Assessment. United Nations Food and Agriculture Organisation.

FAO, 2010. Food and Agriculture Organisation. Global Forest Resource Assessment, FAO, Rome, Italy.

GRZ 2002. Poverty Reduction Strategy Paper, 2002−2005. In: GRZ (Ed.), Lusaka, Zambia.

GRZ 2009. Public Private Partnership Act of 2009. In: MOFNP (Ed.), Lusaka, Zambia.

Guveya, E., 2006. Final Draft Report of the Mid-term Baseline Survey for Forest Resource Management Project (FRMP). Ministry of Tourism, Environment and Natural Resources (MTENR). Forestry Department, Lusaka, Zambia.

Kabananukye, K.I.B., Kabananukye, A.E.K., Krishnamurty, J., Owomugasho, D., 2004. Economic Growth, Employment, Poverty and Pro-Poor Policies. Uganda. Issues in Employment and Poverty. Discussion Paper 16. Recovery and Reconstruction Department. ILO, Geneva.

Margaret, A., Geller, S., Mcconnell, R., Tumushabe, G., 2006. Linking National Forest Programmes and Poverty Reduction Strategies, Report of FAO, FAO Mission report.

Mogaka, H., Gasheke, S., Turpie, J., Emerton, L., Karanja, F., 2001. Economic Aspects of Community Involvement in Sustainable Forest Management in Eastern and Southern Africa. IUCN-Eastern Africa Regional Office. Forest and Social Perspectives in Conservation No. 8. Nairobi, Kenya.

Mulombwa, J., 1998. Non-Wood Forest Products in Zambia. A Report of the EC-FAO Partnership Programme (1998−2000). Project GCP/INT/679/EC Data Collection and Analysis for Sustainable Management in ACP Countries — Linking. National and International Efforts. Forest Department, FAO, Rome, Italy.

Mwitwa, J., German, L., Muimba-Kankolongo, A., Puntodewo, A., 2012. Governance and Sustainability Challenges in Landscapes Shaped by Mining: Mining-Forestry Linkages and Impacts in the Copper Belt of Zambia and the DR Congo. For. Policy Econ. 25, 19−30.

Ng'andwe, P., Muimba-Kankolongo, A., Shakacite, O., Mwitwa, J., 2006. Forest Revenue, Concessions Systems and the Contribution of the Forestry Sector to Zambia's National Economy and Poverty Reduction. FEVCO, Lusaka, Zambia.

Ng'andwe, P., Muimba-Kankolongo, A., Mwitwa, J., 2010. Forestry Sector Guidelines for Data Collection and Handling. Mission Press, Ndola, Zambia, pp. 63.

Nguvulu, C.Z., 2005. Selection Priorities for Wild Fruit Trees in Zambia: The Farmer's Perspective. Master of Science in Agroforestry. University of Wales, UK.

Puurstjärvi, E., Mickels-Kokwe, M., Chakanga, M., 2005. The Contribution of the Forest Sector to the National Economy and Poverty Reduction in Zambia. SAVCOR, Lusaka, Zambia.

Ratnasingam, J., Ng'andwe, P., 2012. Forest Industries Opportunity Study — Synthesis Report Submitted to the Forestry Department Integrated Land Use Assessment II and the Food and Agriculture Organisation (FAO) of the United Nations.

UN, 2003. Integrated Environmental and Economic Accounting. United Nations, European Commission, International Monetary Fund, Organisation for Economic Co-operation and Development and World Bank. New York, USA.

Vinya, R., Syampungani, S., Kasumu, E.C., Monde, C., Kasubika, R., 2012. Preliminary Study on the Drivers of Deforestation and Potential for REDD + in Zambia. A Consultancy Report Prepared for Forestry Department and FAO under the National UN-REDD+ Programme Ministry of Lands & Natural Resources. Lusaka, Zambia.

Whiteman, A., 2004. An Appraisal of the Licensing and Forest Revenue System in Zambia. Ministry of Tourism, Environment and Natural Resources, Lusaka, Zambia.

Whiteman, A., Brown, C., 1999. The potential role of forest plantations in meeting future demands for industrial wood products. Int. Forestry Rev. 1 (3), 143–152.

ZAVC, 2003. Zambia Vulnerability Assessment Committee, Livelihoods and Vulnerability Assessment Final Report, Lusaka, Zambia.

ZFAP, 1997. In: MENR (Ed.), Zambia Forestry Action Plan 1997–2015. Forestry Department, Lusaka, Zambia.

INDEX

Printed in the United States
By Bookmasters